Copper in the Early Modern Sino-Japanese Trade

Monies, Markets, and Finance in East Asia, 1600–1900

Edited by

Hans Ulrich Vogel

VOLUME 7

The titles published in this series are listed at *brill.com/mmf*

Copper in the Early Modern Sino-Japanese Trade

Edited by

Keiko Nagase-Reimer

BRILL

LEIDEN | BOSTON

Cover illustration: From "Scroll with views of the Dutch Factory and Chinese Quarter in Nagasaki 唐館図蘭館図絵巻" drawn by Ishizaki Yūshi 石崎融思. Courtesy of Nagasaki Museum of History and Culture 長崎歴史文化博物館.

Library of Congress Cataloging-in-Publication Data

Names: Nagase-Reimer, Keiko.
Title: Copper in the early modern Sino-Japanese trade / edited by Keiko Nagase-Reimer.
Description: Leiden : Brill, 2016. | Series: Monies, markets, and finance in East Asia, 1600-1900, ISSN 2210-2876 ; volume 7 | Includes bibliographical references and index.
Identifiers: LCCN 2015029107| ISBN 9789004299450 (hardback : acid-free paper) | ISBN 9789004304512 (e-book)
Subjects: LCSH: Copper industry and trade—Japan—History. | Copper industry and trade—China—History. | Japan—Commerce—China—History. | China--Commerce—Japan—History. | Japan—Economic conditions—1600–1868.
Classification: LCC HD9539.C7 J323 2016 | DDC 382/.4566930952—dc23
LC record available at http://lccn.loc.gov/2015029107

This publication has been typeset in the multilingual "Brill" typeface. With over 5,100 characters covering Latin, IPA, Greek, and Cyrillic, this typeface is especially suitable for use in the humanities. For more information, please see www.brill.com/brill-typeface.

ISSN 2210-2876
ISBN 978-90-04-29945-0 (hardback)
ISBN 978-90-04-30451-2 (e-book)

This book is printed on acid-free paper.

Dedicated to
Professor Nakayama Shigeru 中山茂
1928–2014

Contents

Foreword

The project "Monies, Markets and Finance in China and East Asia, 1600–1900," funded by the German Research Foundation (DFG: Deutsche Forschungsgesellschaft) and led by Hans Ulrich Vogel of the Institute for Chinese Studies at the University of Tuebingen, started in 2005. This project included the involvement of scholars not only from the field of Chinese studies, but also of Japanese studies, history of economics, history of technology, and geography in Germany and abroad. As part of this project, the Japanese History Section of the Faculty of East Asian Studies at the Ruhr University Bochum was responsible for researching the production, transportation and export of Japanese copper. This book is one of the results of that research.

When we started the research project, our project team at the Ruhr University Bochum saw it as a great chance to learn more about a whole range of topics: about the history of mining in Japan, the technologies it used and its political and social environment; about the development of a market for minerals and metals in Japan and in neighbouring countries; and about Sino-Japanese trade during the Edo period (1600–1868) and its impact on both sides. Even though research on each of these topics had been done in Japan—and to a much lesser extent in the West—they were mostly treated as separate entities. There was hardly any effort to pursue questions that linked production, market conditions and trade to each other.

In the field of mining history, our main topic during the first phase of the project, this fragmentation became even more apparent. Most books and articles on this topic focus on regions or on individual mines. This has remained the dominant perspective in mining history far into the modern period. Even one of the foremost mining historians in Japan, Kobata Atsushi (1905–2001), rarely attempted to go beyond a regional approach to establish a national or international perspective on Japanese mining. The lack of such a comprehensive mining history probably reflects the state of the written and visual sources: they are scattered among numerous archives, museums, libraries and private holdings. That makes it very difficult to obtain a bigger picture. But it also shows the difficulties that the field faces in Japan if it is to become a research subject on its own, like "Montangeschichte" (mining history) in Germany and in other European countries.

When we started out, we hoped to be able to contribute to the compilation of such a comprehensive mining history. But in the course of our research we realised that the vast amount of sources and other materials would not allow us to implement this idea within the framework of the project. So we focused on various aspects of mining, mining technology, transport, etc. In the project's second phase, we further elaborated the links between production, market conditions and trade within Japan by following the trail of copper from the mines to Nagasaki, and from there to China. At the same time, we widened our perspective by including research by Japanese colleagues on other products traded between China and Japan, such as sugar, dried marine products and books, and by analysing the function of these products and their interaction with copper in the exchange of goods. The articles in this book, including those by Japanese colleagues that appear for the first time in English here, present some of the results and insights gained from these studies. They show that some ideas about the dominating power of the Shogunate and the mechanisms of trade and finance in the Edo period have to be reassessed.

Working with colleagues from Tuebingen, Heidelberg, and various places in China, Japan and worldwide for six years has been an extremely gratifying experience for me and, I am sure, for all those involved in the project. Thanks to the tireless efforts of Hans Ulrich Vogel the various groups and participants remained in close contact with each other throughout the whole period. We all profited greatly from ideas and advice from other members of the research group, but also from external specialists who were invited to our meetings and workshops.

One of those who were a great help during the whole six years was Prof. Nakayama Shigeru, a well-known specialist on the Japanese history of science and technology. A cosmopolitan, he served on the project's advisory board and, despite his busy schedule, came to all our conferences. He was an inspiring and reliable mentor who provided us with valuable advice and support. It was a great shock when we heard of his death in May 2014. We deeply mourn this great scholar, whose death is a loss for us and all his many friends, colleagues and former students all over the world.

This book is dedicated to the memory of Professor Nakayama Shigeru.

Regine Mathias
Bochum, November 2014

Acknowledgements

This book is an outcome of the research carried out at Ruhr University Bochum within the framework of the project entitled "Monies, Markets and Finance in China and East Asia, 1600–1900" that was supported by the German Research Foundation (DFG). Two of the authors, Iwasaki Yoshinori and Matsuura Akira, participated in one of our international workshops sponsored by the DFG, the Baden-Württemberg Association for the Advancement of Cooperation with the People's Republic of China (Baden-Württembergische China-Gesellschaft, Konstanz), the German-East Asian Forum of Sciences and Humanities (Deutsch-Ostasiatisches Wissenschaftsforum, Tübingen), and Universitätsbund Tübingen—Vereinigung der Freunde der Universität Tübingen. These workshops gave scholars the opportunity to exchange ideas and inspiration in a cordial atmosphere. I would like to express my gratitude firstly to Hans Ulrich Vogel and to the scholars of this project, especially Shan Kunqin and Alexander Jost, as well as to all the sponsors of these international workshops and for making them such a success. Liu Shiuh-Feng and Imai Noriko were unfortunately not able to take part in the workshops, but they kindly made their essays available for this volume.

Thanks are also due to Michael Thornton, Karen Finney-Kellerhoff, Sabine Prüfer, Eriko Ogihara-Schuck and Benjamin Stürmer for translating the Japanese essays into English and for proofreading the English texts. In editing this volume, I received a great deal of support from Dagmar Reinders, Pierre Kemper and Takeda Yuhono. And finally, for all the advice that she offered in connection with the editing of this volume, I would like to express my gratitude to Patricia Radder and Peter Buschman at Brill. Without their help this book would not have been possible.

I would also like to thank Regine Mathias, leader of the part project at Ruhr University Bochum, for supporting and advising me on all aspects. It was a great pleasure for me to take on this project under her leadership. Thomas Hirzel of the Institute for Chinese Studies at the University of Tübingen has conducted extensive research on the Sino-Japanese copper trade from the Chinese perspective and has given me valuable insights into this topic. Furthermore, I would like to thank George Bryan Souza of the University of Texas for introducing the concept of the "commodity chain" to our group; Kuroda Akinobu of the Institute for Advanced Studies on Asia at Tokyo University and David Mervart of the Japanese History department at the University of Heidelberg for their lectures at Ruhr University Bochum, and Stefan Dieball for his marvellous cartographies. I am indebted to colleagues at Ruhr University Bochum, the members of our research group and the participating scholars for all their exciting ideas. Last but not least, I would like to express my deep sorrow at the death of Professor Nakayama Shigeru. Since attending an undergraduate course on the history of science and technology I have been familiar with his inspiring works and was thus delighted when he joined this project as a member of the advisory board. I owe him a huge debt of gratitude for his encouragement and support.

Keiko Nagase-Reimer
Bochum, December 2014

Notes

1. Japanese and Chinese names appear in their original order, with the surname first, then the first name.

2. Years have been converted into the Gregorian calendar, while days and months are left in the original numbering. For example, 1689/4/3 stands for the third day of the fourth month of the year 1689. Leap months are marked with "i". For example, 1710/i8/26 stands for the twenty-sixth day of the eighth month (leap month) in 1710.

3. Weights and Measures:
The measures varied by region and time period. Thus it is difficult to give an accurate conversion. Therefore, the original terms are left in the text.

Weights:

1 *jin* (Chinese)/*kin* (Japanese) 斤 = 600 g
1 *tan* 擔[担] = 100 *kin* = 60 kg
1 *ko* 箇 = 100 *kin* = 60 kg
1 *maru* 丸 (Japanese) = 50 *kin* = 30 kg
1 *kan* 貫 = 1,000 *monme* 匁 (Japanese) = 3.75 kg

The unit "*kan* 貫" is practically the same as "*kanme* 貫目."

Volume:

1 *koku* 石 = 180.39 litre

List of Figures, Tables and Maps

List of Figures

List of Tables

List of Maps

Glossary

Akita han 秋田藩	Akita domain
Ani dōzan 阿仁銅山	Ani copper mine
aradō 荒銅	crude copper
Ashio dōzan 足尾銅山	Ashio copper mine
awasebuki 合吹	alloy melting
azana 字	pen name
azuki 小豆	ca. 5 mm long oval-shaped beans growing in Asia. Its colour ranges from reddish-brown to maroon.
bakufu 幕府	Shogunate
Baoquanju 寶泉局	metropolitan mints under the Board of Revenue
Baoyuanju 寶源局	metropolitan mints under the Board of Works
Besshi dōzan 別子銅山	Besshi copper mine
Betsudan shōhō 別段商法	special trade
bugyō 奉行	magistrate
Bunkyū eihō 文久永宝	hundred-*mon* copper coins in circulation in the late Edo period
busshukan 佛手柑	fingered citron
chiuridō 地売銅	copper for domestic use
chōgin 丁銀	various silver coins in circulation from the 14th century to the end of the Shogunate in Japan
chōkō 朝貢	tribute trade
Chūyō 中庸	*The Doctrine of the Mean*
daibutsu 大仏	a large statue of the Buddha
Daigaku 大学	*The Great Learning*
daiku 大工	Smelter workers
daimyō 大名	feudal lords
deiriunjōgin 出入運上銀	tax levied on ships arriving in port
Dejimajiro 出島白	Dejima White (sugar)
Detai 得泰	name of a Chinese ship
dōfukiya 銅吹屋	copper refinery
dōfukiya nakama 銅吹屋仲間	association of refineries
dongren 東人	Easterners
dōtoiya / dōton'ya 銅問屋	copper merchant house
dōya 銅屋	traditional copper merchants
dōza 銅座	Copper Agency
Ekikyō 易経	*The Classic of Changes*
eshang 額商	a limited number of private merchants, "quota merchants"

fuchimai 扶持米	rice allowance
fukahire 鱶鰭	shark fin
fukidoko 吹床	furnace room
funatekazari 船手錺	a type of simultaneous decoration and hull reinforcement used on Japanese ships
fuyu 鰒魚	dried abalone
gaitu guiliu 改土歸流	circulating officials
Genbun dōza 元文銅座	Second Copper Agency
Genroku dōza 元禄銅座	First Copper Agency
geshi 夏至	10th solar term, summer solstice
gezai 下財	low-paid workers
gokyō'ōkata 御饗応方	banquet expenses
gotomoyuki no uchi 御知行之内	salary for *kerai*
gotōyōkata 御当用方	daily expenses in the Edo *hantei*
goyōdō 御用銅	copper for export (literal translation: copper for official use)
gozenkin 御前金	money used by a feudal lord
guanchai 關差	internal customs station superintendent
guanshang 官商	state merchants
guotanghang 過塘行	transport workers
haibuki 灰吹	cupellation or ash smelting
haifukijōgin 灰吹上銀	silver obtained through cupellation
haishen 海參	dried sea cucumber
hanmai 飯米	rice for daily consumption
hantei 藩邸	residence maintained by a *daimyō*
hao 耗	compensatory copper
hiyamizuuri 冷水売	iced water vendors
hong cai 紅菜	cockscomb algae
hoshiawabi 干鮑	dried abalone
ichimonsen 一文銭	one-*mon* coins
Igakukan 醫學館	a medical school founded by the Shogunate
iriko 煎海鼠	sea cucumber
Iwami ginzan 石見銀山	Iwami silver mine
jikiyama 直山	mine under direct management
jiuqing 九卿	the highest-ranking officials
jōdaka 定高	trade ceiling
jōdakasei 定高制	trade law implemented in 1685
jōhaku 上白	a type of sugar
kaisendon'ya 廻船問屋	a kind of shipping agent
kaishi 回賜	gifts

kaisho 会所	clearing house for foreign trade accounts
kaiukemai 買請米	rice purchasing system
kanahori 金堀	miners
kanako かなこ	managers of a small mining area
Kan'ei tsūhō 寛永通宝	a name given to coins in circulation from 1636 to the end of the Shogunate
kanpō 漢方	a form of traditional Chinese medicine
kansho 甘蔗	sugar cane
karō 家老	senior retainer
kashiwamochi 柏餅	bean jam rice cakes
katsuobushi 鰹節	bonito flakes
keichō gin 慶長銀	Keichō silver coin
kerai 家来	retainers
Kinton 金團	mashed sweet potatoes
Kinzanji miso 金山寺味噌	Kinzanji fermented soybean paste
koban 小判	an oval gold coin
kobuki 小吹	process in which molten copper was poured from the crucible into a mold that had been placed in water
kodō 古銅	scrap copper
kogan 古玩	old books of particular rarity or value
konazatō 粉砂糖	powdered sugar
konpeitō 金平糖	candy
koorizatō 氷砂糖	rock candy
koshi 古詩	ancient poems
kubutsu 公物	public property
kuramai 蔵米	stored rice
kuramoto 蔵元	merchants who handled domains' rice and special products in the Edo period
kurozatō 黒砂糖	muscovado (a type of unrefined brown sugar)
kyūchimai 給知米	payroll rice to be supplied to a retainer
kyūshō bōeki 求償貿易	compensation trade
mabuki 間吹	process in which ore is heated in a solution to extract impurities
machi toshiyori 町年寄	city elders
manjū 饅頭	steamed buns
mashidō 増銅	additionally-sold copper
myōdai 名代	agent
Meirindō 明倫堂	Meirin Temple
Meiwa dōza 明和銅座	Third Copper Agency
mikan 蜜柑	tangerines

miso 味噌	fermented soybean paste
Mitsuishi konbu 三ツ石昆布	*Mitsuishi* kelp, Saccharina angustata
mononarimai 物成米	annual rice tax
Mōshi 孟子	*The Mencius*
motokata bugyō 本方奉行	finance magistrate
motone 元値	cost price
na 名	given name
Nagasaki kaisho 長崎会所	Nagasaki clearing house for foreign trade accounts
namarinagashidoko 鉛流床	areas for lead refining
nanbanbuki 南蛮吹	Saiger process
Nanbuton'ya 南部問屋	trading houses handling Nanbu products
Nara Tōdaiji 奈良東大寺	Tōdai Temple in Nara
neiwufu shangren 内務府商人	Imperial Household Department merchants
nengumai 年貢米	annual rice tax
neriyōkan 練り羊羹	jellied bean paste
ninjin 人参	ginseng
obon お盆	a festival
okidemai 沖出米	rice to be sold to other domains
okideyakugin 沖出役銀	tax levied on the import and export of materials at ports within the Akita domain
osadamedaka 御定高	trade ceiling
Osaka ton'ya 大坂問屋	Osaka merchant houses
Osarizawa dōzan 尾去沢銅山	Osarizawa mine
Ōsugimaru 大杉丸	(name of a Japanese ship)
qian 錢	copper cash
qianjie ling 遷界令	Coastal Evacuation Order
qianju 錢局	mints
Qinding Da Qing huidian 欽定大清會典	*Collected Statutes of the Great Qing*
Raiki 礼記	*The Classic of Rites*
rakusatsune 落札値	sale price
risshi 律詩	ancient Chinese verse form
Rongo 論語	*The Analects*
rushi 留師	craftsmen who made crucibles
ryō 両	a gold currency unit
ryōshu 領主	local lords
sakiurimai 先売米	rice for advance payment
sakoku seisaku 鎖国政策	seclusion policy

sanbon 三盆	fine sugar
sanka seido 三貨制度	trimetallic coinage system introduced by the Shogunate (literal translation: three coin system)
saodō 棹銅	refined copper bars
satōiri kintoki 砂糖入金時	sugared beans
satōmizuuri 砂糖水売	sugar water vendor
sekifu 石譜	genealogies and suchlike, printed using a stone plate
sengoku jidai 戦国時代	Warring States Period (1467–1573)
sentō 泉糖	type of sugar
shiboribuki 鉸吹	liquidation, or the method known as *nanbanbuki* in a narrow sense
shierjia 十二家	another name for the quota merchants, Twelve Families
Shimazu *han* 島津藩	Shimazu domain
Shikyō 詩経	*The Classic of Poetry*
shinpai 信牌	tallies or permits for trading at Nagasaki, Japanese Trade License
shiratamauri 白玉	rice flour dumplings
shiromonogae 代物替	barter trade. Import goods were not paid with gold or silver, but with copper, *tawaramono, shoshiki* and others.
shirozatō 白砂糖	white sugar
shōga 生薑	ginger
shōgun 将軍	feudal military administrator
Shokyō 書経	*The Classic of History*
shōsetsu 小雪	20th solar term
shoshiki 諸色	miscellany
shōsho 小暑	11th solar term
Shōtoku shinrei 正徳新例	Shōtoku New Law, Shōtoku Regulations
shukusha 宿砂	cardamom
Shunjū 春秋	*The Spring and Autumn Annals*
Shushigaku 朱子学	Neo-Confucianism
soboku 蘇木	sappan wood
suichū chūzō 水中鋳造	in-water casting
tang-guan 唐館	Chinese quarters (Chinese)
taihaku 太白	type of sugar
taisetsu 大雪	21st solar term
tawaramono tehon 俵物手本	samples of dried goods
teategin 手当銀	subsidy
tedai 手代	clerk

tedai gashira 手代頭	manager
tengusa てん草	seaweed
tenmondō 天門冬	dried asparagus root
Tenpō tsūho 天保通宝	four-*mon* copper coins
tenryō 天領	territory of the Shogunate
tōgan 冬瓜	winter melon
Tōjin yashiki 唐人屋敷	Chinese quarters (Japanese)
tokoya 床屋	smelting station
tosaka 鶏冠	seaweed /cockscomb
Tōsen 唐船	Chinese merchant sailing ships
ukeyama 受山	mine under contract management
unjōkin 運上金	business tax
Wagōmaru 和合丸	name of a Japanese ship
yakigama 焼釜	roasting kiln
yamashi 山師	mine operator, mining contractor
Yangmao 永茂	name of a Chinese junk
yanwo 燕窩	swallow's nest
yasumiyama 休山	mine not currently in operation
yodō 余銅	left-over copper
yoriki 与力	administrative assistant in the Shogunate's office
yuchi 魚翅	shark fin
yurimonoshi 汰りもの師	craftsmen who collected copper from the smelting rooms
zeni 銭	copper cash
zhanhailing 展海令	Order of the Open Seas
zhiqian 制錢	official standard copper cash

About the Contributors

Imai Noriko 今井典子
specializes in the mining history of early modern Japan. She took her BA and MA in Japanese history at Tokyo University and was research scholar (1978–2000) and vice director (2001–2005) of the Sumitomo Historical Archives (Sumitomo shiryōkan 住友史料館). In her work for the archive she was engaged in re-publications of historical material related to the Sumitomo family and its business and edited eleven volumes of "Sumitomo History Library". Her publications include "Kinsei Nihon no dō: dōshijō to dōtōsei" 近世日本の銅—銅市場と銅統制—[Copper in Early Modern Japan: Copper Market and Copper Monopoly] (in *Sumitomo shiryōkanpō* [Journal of Sumitomo Archives] 39, 2008) and "Kinsei Sumitomo no fukisho no kenkyū" 近世住友の吹所の研究 [Research on the Sumitomo refinery in the early modern era] (in *Sen'oku sōkō*, vol. 19, 1980).

Iwasaki Yoshinori 岩崎義則
is assistant professor at the Department of Japanese History of Kyūshū University. He received his Ph.D. from Kyūshū University in 1997 with the thesis *Kinsei dōza no kenkyū* 近世銅座の研究 [Study on Copper Agencies in early modern Japan] under the supervision of Prof. Nakamura Tadashi. His published research focuses on copper production and transport in early modern Japan. His current interests are the cultural activities of the *daimyō* Matsura Family in Hirado during the Edo period. His publications include *Higashi Ajia sekai no kōryū to henyō* 東アジア世界の交流と変容 [Intercommunication and transformation in East Asia] (Fukuoka: Kyūshū daigaku shuppan, 2011), co-edited with Morihira Masahiko and Takayama Michiaki, and *Nagasaki ken nai no tayōna shūraku ga keisei suru bunkateki keikan hozon chōsa hōkokusho, shiryō hen 3* 長崎県内の多様な集落が形成する文化的景観保存調査報告書 資料編3 [Report on the preservation of a cultural landscape formed by various settlements in the Nagasaki prefecture] (Nagasaki: Nagasaki ken, 2013).

Keiko Nagase-Reimer 永瀬ライマー桂子
specializes in the history of technology in Japan. She received a BA in instrumentation engineering from the Faculty of Science and Technology of Keio University, an MA in history of science and technology as well as in physics from the Technical University of Berlin, and a Ph.D. in history of science and technology from the Technical University of Berlin. From 2005 to 2011 she concentrated on copper production and trade in Tokugawa Japan within the

framework of the DFG project "Monies, Markets, and Finance in East Asia, 1600–1900." Her publications include *Mining, Monies, and Culture in Early Modern Societies* (Leiden: Brill, 2013), co-edited with Nanny Kim, and *Forschungen zur Nutzung der Kernenergie in Japan, 1938–1945* [Research on application of atomic energy in Japan, 1938–1945] (Marburg: Förderverein Marburger Japan-Reihe, 2002).

Liu Shiuh-Feng 劉序楓

is an Associate Research Fellow at the Research Center for Humanities and Social Sciences, Academia Sinica. He received his Ph.D. in history from Kyūsyū University. He specializes in Sino-Japanese relations and East Asian maritime history. In recent years, his research interests have focused on the history of shipwrecks in East Asia, including China, Japan, Ryūkyū and Korea. His publications include *Qing dai dang an zhong de hai nan shi liao mu lu- She wai pian* 清代檔案中的海難史料目録一涉外編 [Catalogue of Shipwreck Records in the Ch'ing Archives: Foreign-Related Materials] (Taipei: Academia Sinica, 2004) and "Shipwreck Salvage and Survivors' Repatriation Networks of the East Asian Rim in Qing Dynasty" (in Fujita Kayoko, Momoki Shiro, Anthony Reid (eds.), *Offshore Asia: Maritime Interactions in Eastern Asia Before Steamships*, pp. 211–235, Singapore: Institute of Southeast Asian Studies, 2013).

Matsuura Akira 松浦章

in Japanese history and in cultural interaction studies, is professor at the Faculty of Letters, Kansai University and director of the Center for the Study of Asian Cultures, Kansai University. He has been engaged in research on the history of transportation during the Ming and Qing Dynasties, especially on commodity transport by Chinese junks. Recently he has turned to the cultural exchange among countries surrounding the East China Sea during the Ming and Qing Dynasties. His publications include *Edo jidai tōsen ni yoru Nicchū bunka kōryū* 江戸時代唐船による日中文化交流 [*Sino-Japanese cultural interactions via the Edo-period Chinese junk trade*] (Kyoto: Shibunkaku shuppan, 2007), *Shindai hosen enkai kōunshi no kenkyū* 清代帆船沿海航運史の研究 [Study on the history of coastal journeys by junks during the Qing Dynasty] (Osaka: Kansai daigaku shuppanbu, 2010) and *Kinsei Higashi Ajia kaiiki no bunka kōshō* 近世東アジア海域の文化交渉 [*Cultural interactions in the early modern East Asian maritime region*] (Kyoto: Shibunkaku shuppan, 2010). His most recent publication is *Kisen no jidai—kindai higashi Ajia kaiiki* 汽船の時代 近代東アジア海域 [Age of steam vessels—early modern East Asian maritime region], (Osaka: Seibundō shuppan, 2013).

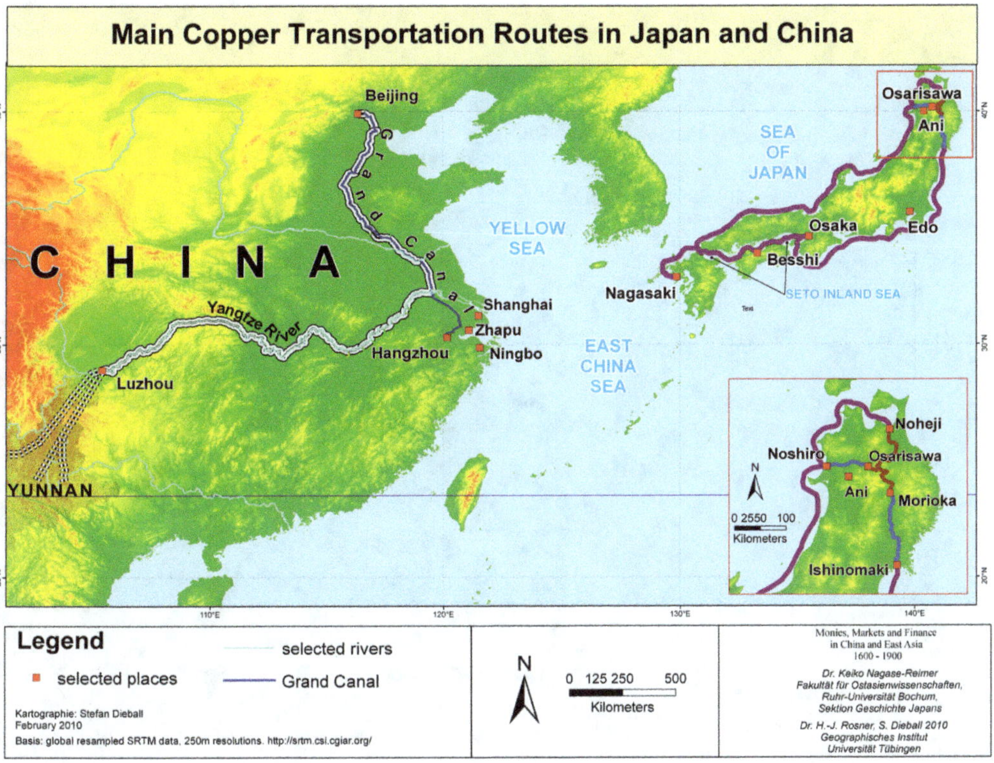

Main Copper Transportation Routes in Japan and China

CHAPTER 1

Introduction

Keiko Nagase-Reimer

Mining began relatively late in Japan. The first mining boom started in the middle of the 7th century. Until then, Japan had been dependent on imported refined metals from Korea. During this first boom, copper was produced in much larger quantities than gold or silver and was used to make coins and ceremonial objects. Most famously it was used for the construction of the Great Buddha statue in Nara (747–752). This mining boom lasted until the 9th century.[1]

The history of copper in Sino-Japanese trade dates back to the 10th century when Japan ceased producing coins. At that time Chinese copper coins became the medium of exchange for expanding market activities and the demand for them in Japan was high. From the early 12th century, however, China began to suffer a shortage of copper, a situation that continued until the 18th century. Against this background, the Chinese government attempted to ban the export of coins to Japan. However, Japanese merchants imported increasing quantities of Chinese copper coins in the late 12th and 13th centuries.[2] Finally, when the Kamakura Shogunate (*Kamakura bakufu* 鎌倉幕府 1185–1333) in Japan collapsed and the Yuan Dynasty 元朝 (1279–1368) rose to power in China, Sino-Japanese trade came to a standstill and consequently the trade in copper also ceased.[3]

Trade between the two countries re-started when political stability was restored by the Muromachi Shogunate (*Muromachi bakufu* 室町幕府 1336–1568) in Japan and the Ming Dynasty (1368–1644) in China. It was driven by the demand for goods like silk and silver and copper coins in Japan. Sino-Japanese trade in the 15th century was mostly conducted under the tributary system, but some private trade was also permitted subject to strict control by the Chinese.[4]

The second mining boom in Japan began in the mid-15th century and continued for about a century. During the so-called Warring States Period (*Sengoku jidai* 戦国時代 1467–1573) mining and refining activities intensified due to the

1 Murakami, *Kin gin dō no Nihonshi*, 52–54; Kobata, *Kōzan no rekishi*, 5–6.

2 Glahn, *Fountain of Fortune*, 53–54.

3 Ibid., 89.

4 Ibid., 90–91.

 | DOI 10.1163/9789004304512_002

increased quantity of gold and silver in circulation, which was used as reward and payment for soldiers. Because the feudal lords (*sengoku daimyō* 戦国大名) required these metals in order to maintain their armies, they were eager to develop gold and silver mines in their territories.[5] Technical innovations in the field of mining and metallurgy from the end of the 15th century to the beginning of the 16th century, such as, for example, the improvement of surveying techniques, the production of iron tools and the introduction from Korea of a new refining process called cupellation, increased productivity.[6] The demand for silver grew not only within Japan, but also outside the country. Accordingly, it became one of the main export goods during the 16th century in Japan and was mostly exchanged for silk.[7] At the beginning of the 17th century, Japan's silver exports amounted to two hundred tons annually, which was one-third of world silver output.[8] During this period, which Flynn and Giráldez called "the Potosi/Japan silver cycle,"[9] Japan was engaged in both regional and global trade.

In 1636 Japan introduced the so-called Seclusion Policy (*sakoku* 鎖国). The Tokugawa Shogunate, which had ruled Japan since 1603, prohibited Japanese from travelling overseas. Furthermore, only Chinese and Dutch ships were permitted to call at Nagasaki.[10] As Ronald Toby[11] and Tashiro Kazui 田代和生[12] observed, however, Nagasaki was not the only port where the exchange of commodities with foreign countries took place. Foreign trade was also conducted through three other routes: Tsushima 対馬 served as a gateway to Korea, Satsuma 薩摩 traded with Ryūkyū 琉球, and Matsumae 松前 had contacts with Russia.

In the late 17th century, when the silver mines were exhausted, the feudal lords promoted copper mining in their territories and copper production increased accordingly. At the turn of the century, copper production in Japan rose to its peak level of ca. 6,000 tons. Of this, ca. 5,400 tons were exported, and from then on copper replaced silver as the main export commodity. The Chinese state was the chief buyer of Japanese copper, as shown in figure 1.1.

---

5 Kobata, *Nihon dōkōgyōshi no kenkyū*, 3; Murakami, *Kin gin dō no Nihonshi*, 52 and 129.

6 Keiko Nagase-Reimer "Water Drainage in the Mines in Tokugawa Japan: Technological Improvements and Economic Limitations," 26.

7 Arano, *Sakoku no kenkyū*, 6 and 8.

8 Kobata, *Kōzan no rekishi*, 59; Arano, *Sakoku no kenkyū*, 8.

9 Flynn and Giráldez, "Cycles of Silver: Global Economic Unity through the Mid-Eighteenth Century," 399.

10 Arano, *Sakoku no kenkyū*, 12–13.

11 Toby, "*Sakoku*" *to iu gaikō*, 12–20.

12 Tashiro, "Foreign Relations during the Edo Period: Sakoku Reexamined," 283–306.

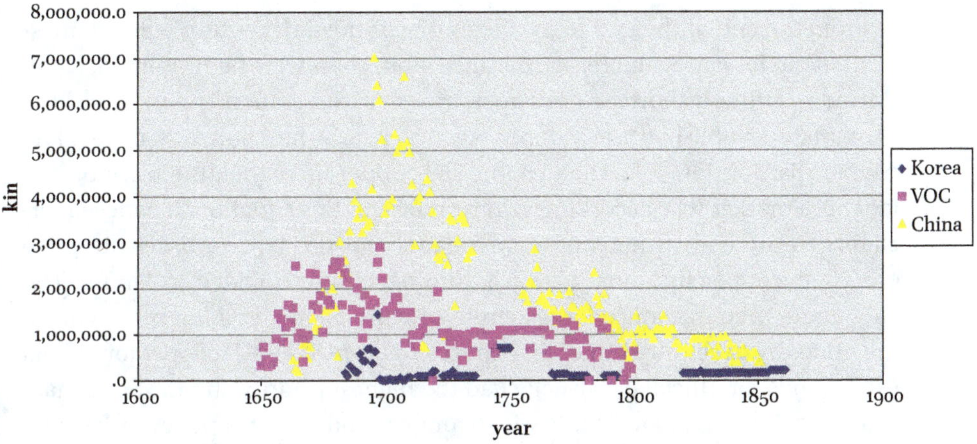

FIGURE 1.1 *Japan's copper exports.*[13]

Chinese merchants procured the copper in Nagasaki and transported it to China under the control of Chinese officials. At the end of the 17th century, the requirements of China's mints were met almost entirely with Japanese copper. When the copper mines in the southeastern province of Yunnan 雲南 began large-scale production in the 1720s, the demand for Japanese copper in China started to fall. But still the mint in the Lower Yangzi region, located on China's east coast, continued to manufacture coins using Japanese copper.[14] Also for the Dutch East India Company (VOC: *Vereenigde Oostindische Compagnie*), another main trade partner of Japan besides the Chinese state, copper became their main import good when the Shogunate prohibited the sale of silver to the VOC in 1668.[15] They generated substantial income from the sale of Japanese copper in South Asia, Europe and the Middle East.[16] However, copper exports via both the VOC and the Chinese merchants declined as the Shogunate tightened its control on copper export after 1715.[17] Japanese copper was also exported to Korea via Tsushima, though the quantity was generally limited to 100,000 *kin*[18] and was thus far smaller than the amount traded at Nagasaki.

13 Tashiro, "Tsushima-han no Chōsen yushutsu dō chōtatu ni tsuite," 149, 164, 178, 184, 186, 190; Kobata, *Nihon kōgyōshi no kenkyū*, 659ff; Shimada, *The intra-Asian trade in Japanese copper by the Dutch East Company during the eighteenth century*, 198.

14 Yamawaki, *Nagasaki no Tōjin bōeki*, 176–77.

15 Shimada, *The Intra-Asian Trade*, 14.

16 Ibid., 37–44.

17 Ishii "Bakumatsu ni okeru bakufu no dō yushutsu kinshi seisaku," 1–2; Yamawaki, *Nagasaki no Tōjin bōeki*, 140–41.

18 Tashiro, "Tsushima-han no Chōsen yushutsu dō chōtatsu ni tsuite—bakufu no dō-tōsei to Nissen dō-bōeki no suitai," 164–69.

From the end of the 17th to the beginning of the 18th century, when domestic demand for copper increased because more coins were being minted, Japan started to suffer a shortage of copper. In order to maintain export levels, the Shogunate established the Copper Agency whose function was to regulate copper distribution.[19] As the demand for copper increased, the price on the Japanese market rose accordingly. In response to this situation, Japanese merchants negotiated with Chinese merchants and the VOC to bring the price of export copper into line. However, the negotiations ended in failure. The Shogunate then introduced a system to procure copper at a lower price from the mines by declaring copper for export as *goyōdō* 御用銅 (copper for official use). Goods for official use usually had the highest priority and the Shogunate could set their price low.[20] It is widely understood that this policy of keeping the copper price low heavily suppressed copper production in early modern Japan.[21]

Previous Research and the Aim of this Book

This book focuses on Japanese copper exports to China in the Tokugawa period. It tries to give the reader a comprehensive picture of the "life" of the copper produced in early modern Japan: how was copper produced and distributed within Japan, and how was it exported to China and consumed there. Furthermore, other significant import/export goods to and from China are also discussed in order to better understand Sino-Japanese trade in general.

Many scholars have researched the history of Japanese copper from different perspectives:

Regarding copper production in early modern Japan, the research by Japanese scholars has been prolific, whereas the topic has been largely neglected in Western research on Japan. A leading scholar in this field, Kobata Atsushi 小葉田淳, has conducted a wide range of research since the 1950s on the development and management of individual copper mines, using primary sources.[22] Although a large number of detailed studies on individual mines

19 Imai, "Kinsei chūki no chiuridō ni tsuite," 50–51.

20 Imai, "Kinsei Nihon no dō," 24; Sumitomo shūshishitsu, *Dai ichi ji dōza to Sumitomo*, 71; Nagazumi, "Osaka dōza," 410.

21 For example, Ishii, "Bakumatsu ni okeru bakufu no dō yushutsu kinshi seisaku," 3.

22 For example, Kobata, *Nihon kōzanshi no kenkyū*.

have been published, only little attempt[23] has been made to integrate the particular aspects of individual mines into a broader view on the role of copper mining and mining in general in Japanese history.

The literature on mining technology in early modern Japan in general is not abundant, though some studies have been published by Nagahara Keiji 永原慶二 and Yamaguchi Keiji 山口啓二,[24] Nihon kagakushi gakkai,[25] Kobata[26] and Sasaki Jun'nosuke 佐々木潤之介.[27] On the subject of Sumitomo refinery and refining techniques, intensive research has been conducted by Ueda Kōichi 植田晃一[28] and Imai Noriko 今井典子,[29] using rich primary sources in the Sumitomo Archives.

Regarding the Shogunate's policy on copper and the copper market, studies are few, but some valuable research has been done, for example, by Ishii Takashi 石井孝 and Imai Noriko.[30] Of considerable importance is Imai's 2008 publication in which she developed her new interpretation of the role of the Shogunate.[31] With regard to the role of the Copper Agency (*dōza* 銅座), Nagazumi Yōko 長積洋子,[32] Ōta Katsuya 太田勝也[33] and Iwasaki Yoshinori 岩崎 義則[34] have conducted significant research.

Sasaki Jun'nosuke was the first scholar to address the relationship between copper mines and the capital Osaka.[35] Especially on the relation between the Besshi copper mine and Sumitomo capital, Sueoka Teruaki 末岡照啓 and Yasukuni Ryōichi 安国良一 have published valuable studies.[36]

---

23 Only a short overview exists compared to a substantial volume of research on individual mines. For example, Ibid., 3–44.

24 Nagahara and Yamaguchi, *Kōza Nihon gijutsu no shakaishi (Dai go kan): Saikō to yakin.*

25 Nihon kagakushi gakkai, *Saikō yakin gijutsu.*

26 Kobata, *Nihon kōzanshi no kenkyū*, 45–73.

27 Sasaki Jun'nosuke, *Zairai gijutsu no hakken to kinsei shakai. Gijutsu no shakai-shi dai 2 kan.*

28 For example, Ueda, "Edo-jidai no kin, gin, dō seiren gijutsu—dai ikkai Bunken to gijustu no gaiyō," 55–59.

29 For example, Imai, "Kinsei Sumitomo no fukisho no kenkyū."

30 For example, Ishii, "Bakumatsu ni okeru bakufu no dō yushutsu kinshi seisaku," 1–32; Imai, "Kinsei chūki no chiuridō ni tsuite," 50–67.

31 Imai, "Kinsei Nihon no dō: dōshijō to dōtōsei," 13–65.

32 Nagazumi, "Ōsaka dōza," 408–419.

33 Ōta, "Dōza," 157–158.

34 Iwasaki, *Kinsei dōza no kenkyū.*

35 For example, Sasaki, "Kinsei sandō seisaku ni kansuru ichikōsatsu" and Sasaki, "Osaka dōton'ya ni tsuite no oboegaki."

36 For example, Sueoka, "Kinsei zen, chūki ni okeru Sumitomo no keiei kōzō (1)—Toshi shōgyō seiritsu, hatten-ki no kinsei sandō shihon," 1–89; Yasukuni, "Besshi dōzan no son'eki to Sen'oku Osaka honten," 39–96.

Research on Nagasaki trade in general started already at the end of the 19th century. Important work was done by Yamawaki Teijirō 山脇悌二郎[37] and Nakamura Tadashi 中村質[38] on the subject of the trade system and by Ōba Osamu 大庭脩[39] and Matsuura Akira 松浦章[40] on cultural aspects. The Sino-Japanese copper trade has been the subject of a number of studies since the 1920s,[41] and Kobata[42] and Ōta Katsuya 太田勝也[43] continued to develop this work further after WWII. A small number of English-language publications, for example, by John Hall,[44] Helen Dunstan,[45] von Glahn[46] and Bettina Gramlich-Oka[47] have provided valuable insights. The Sino-Japanese copper trade has also been studied from specific perspectives: Robert Innes and Byung Nam Yoon, for example, have conducted studies on the relationship between the copper trade and the development of the Akita mines.[48] And Imai has carried out her research since 2001 from the perspective of Sumitomo and the copper trade.[49] Furthermore, Imai and Iwasaki have conducted studies since 2000 from the perspective of copper administration.[50]

This book adds two new dimensions to this previous research: Firstly, it provides an overview in English of the role of Japanese copper in early modern Sino-Japanese trade. The scarcity of English-language publications has made it difficult for a reader not familiar with Japanese to keep up with the research on the subject. Furthermore, publications in Japanese, which give an overview of this subject, are rare. This book fulfills these desiderata. Secondly, it starts from the latest research results in the field. As mentioned above, a new

37 For example, Yamawaki, *Nagasaki no Tōjin bōeki.*

38 For example, Nakamura *Kinsei Nagasaki bōeki shi no kenkyū.*

39 For example, Ōba, *Edo jidai ni okeru Chūgoku bunka juyō no kenkyū.*

40 For example, Matsuura, *Edo jidai Tōsen ni yoru Nitchū bunka kōryū.*

41 Iwao, "Edō jidai ni okeru dōsen no kaigai yushutsu ni tsuite," 99–110.

42 For example, Kobata, "17seiki narabini 18seiki shoki ni okeru Chūgoku-sen, Oranda-sen ni taisuru wagakuni no dō-yushutsu," 3–6.

43 For example, Ōta, "Dōshiromonogae bōeki no sōshi ni tsuite."

44 Hall, "Notes on the early Ch'ing copper trade with Japan," 444–461.

45 Dunstan, "Safely supping with the Devil: The Qing State and its Merchant Suppliers of Copper," 42–81.

46 Glahn, *Fountain of Fortune.*

47 Gramlich-Oka, "Shogunal Administration of Copper in the Mid-Tokugawa Period (1670–1720)."

48 Innes, *The Door Ajar: Japan's Foreign Trade in the Seventeenth Century*; Yoon, *Domain and Bakufu in Tokugawa Japan: The Copper Trade and Development of Akita Domain Mines.*

49 For example, Imai, "Hōei, Shōtoku-ki no dōbōeki to Sumitomo," 1–38.

50 For example, Imai, "Nagasaki bōeki taisei to Genbun dōza," 59–84; Iwasaki, "Kampō Kanei-ki no bōeki kaikaku to dō tōseisaku," 39–68.

non-traditional interpretation of the role of the Shogunate appeared since we started our project on copper in early modern Japan in 2005: Until recently, the Shogunate—through the Copper Agency—was seen to have created a copper monopoly through the strict control of the distribution of copper.[51] It was generally believed that copper production was suppressed because of the Shogunate's policy of keeping the copper price low.[52] Contrary to this interpretation, it is now suggested that the Shogunate was not as oppressive or coercive as previously thought.[53] Several chapters in this book develop their argument based on this latest view. Accordingly, this volume brings a fresh perspective to the subject.

Organization of this Book

This book comprises seven articles. It begins with Imai Noriko's "Copper in Edo-Period Japan," a comprehensive look at copper production, transport and trade in Japan. This is probably the first publication in English to give such a broad overview of this subject. Japanese copper was mainly exported, but it was also used in Japan in construction, the production of household utensils and for casting coins. Imai's essay illustrates how export, domestic use, market prices and the Shogunate's control of copper changed as copper exports declined.

The second essay is Iwasaki Yoshinori's "The Akita domain and Osaka merchant houses at the time of the establishment of the Meiwa Copper Agency." Taking the example of the Akita domain—one of Japan's largest producers of copper for export—this essay presents a detailed analysis of the exchange of funds and rice between the Akita domain and Osaka copper merchant houses from the 1760s onwards. As mentioned above, a few studies have been done on the relationship between copper mines and capitalists in Osaka which were at the same time copper refineries. This paper is innovative in that it analyzes the relation between a copper mine and a *kuramoto* 蔵元 (merchant who handled rice and special products from the domain) which was not a refinery. In the case of the Akita domain the copper merchant house, which also acted as *kuramoto*, compensated the financial losses of the domain. Originally, the copper merchant houses had provided the domain with monetary advances,

51 For example, Nagazumi, "Osaka dōza," 418–19; Yamawaki, *Nagasaki no Tōjin bōeki*, 174.

52 For example, Ishii, "Bakumatsu ni okeru bakufu no dō yushutsu kinshi seisaku," 3.

53 Historian Imai Noriko first published this view in her paper in 2008. See Imai, "Kinsei Nihon no dō," 57–58.

using the domain's rice production of the following year as collateral. However, as copper became a kind of monopoly product of the Shogunate, the domain began taking loans from the copper merchant houses as advance payment for the copper production of the following year. Given the fact that copper thus became a valuable collateral, Iwasaki argues that its continuously low price was not simply in the interest of the Shogunate, but also in the interest of the copper merchant houses.

Following this, Keiko Nagase-Reimer investigates the question of why the price of export copper also remained low in her essay "There's Method in the Madness." So far historians have assumed that the mines and/or the consumers in Japan indirectly compensated the loss made by selling copper to foreign merchants at low prices.[54] Based on an analysis of the distribution of both copper and foreign goods imported from China, the author argues that there are other plausible explanations for the low copper price than those previously accepted.

How high was the demand for Japanese copper in China? For what purposes was Japanese copper used? How did China procure Japanese copper produced under the circumstances explained in the first three articles? Liu Shiuhfeng's article "Copper administration reform and copper imports from Japan in the Qianlong reign of the Qing Dynasty" gives us an overview of the copper administration system in Qing China, especially during the Kangxi reign (1662–1722) and Qianlong reign (1736–1795), with particular focus on imported copper from Japan. Liu analyzes the copper policy reforms impacting on the import of Japanese copper and describes the role that Japanese copper played in China. Furthermore, he describes the difficulties faced by the Chinese copper merchants: For them, the copper business itself was not very profitable. Here we find a number of similarities between the situations in China and Japan.

The above essays discuss the "life" of Japanese copper in the early modern period, from its production to its transportation to China. Now, we expand our view to other goods traded between China and Japan in order to gain a better understanding of Sino-Japanese trade in general. The three essays that follow are by Matsuura Akira, a specialist on Sino-Japanese cultural and commercial exchange in the Edo period.

The commodities that were traded at Nagasaki changed throughout the Edo period. By the 1680s, copper had replaced silver as the principal export good from Japan to China. However, as from the beginning of the 18th century, Japan's

54 Nagazumi, "Osaka dōza," 418–49; Yamaguchi, "Kingin-zan no gijutsu to shakai," 160–61; Sasaki, "Dōzan no keiei to gijutsu," 184–85; Shimada, *The intra-Asian trade in Japanese copper by the Dutch East Company during the eighteenth century*, 57–59.

copper exports decreased, and by the beginning of the 19th century, dried sea foods such as sea cucumber, abalone and shark fin, which were referred to as *tawaramono* 俵物 (goods packed in straw bags), had taken the place of copper as Japan's major export to China.[55] In his essay "The trade in dried marine products from Nagasaki to China during the Edo period", Matsuura draws on various primary sources and gives a detailed account of how these dried marine products became the principal export good and how demand for them grew in China.

The goods that were imported to Japan from China likewise changed over time. In the 18th century, Japan's main import goods were silk thread, textiles and sugar, but by the beginning of the 19th century, medicine had taken their place.[56] This was due to advances in domestic silk and sugar production. Thus, Japan's economic dependence on imports gradually decreased. Matsuura's essay "The Import of Chinese Sugar in the Nagasaki Junk Trade and its Impact" discusses this phenomenon with a detailed analysis of sugar as an example. Additionally, it considers the huge impact of imported Chinese sugar on Japanese culinary culture.

Finally, in "The Import and Export of Books by Chinese Junks in the Edo Period," Matsuura looks at the books that were at the centre of the Sino-Japanese cultural and academic exchange. Books were traded in both directions, from China to Japan and vice versa. Under the so-called Seclusion Policy there was intense demand for foreign books in Japan, even at exorbitant prices. Such books were then reprinted and distributed domestically. There were even cases where Chinese books were exported to Japan and later shipped back to China as reprints after the original texts had been lost, thus reintroducing their ideas to Qing Dynasty scientific circles. Sino-Japanese trade was therefore not limited to the exchange of physical goods, but also included a considerable amount of scientific and cultural exchange.

It is hoped that the latest research results presented in this book will open new perspectives and thereby encourage fruitful discussion on this topic.

---

55 Nagasaki-ken shi hensan iinkai, *Nagasakiken-shi shiryōhen* 4, 347 and 383; Yamawaki, *Nagasaki no Tōjin bōeki*, 227.

56 Cf. Yamawaki, *Nagasaki no Tōjin bōeki*, 108–9, 197–99.

CHAPTER 2

Copper in Edo-Period Japan

Imai Noriko 今井典子

1 Introduction

Prior to the industrial revolution, Japan was blessed with an abundance of copper. The eighth century was a golden age of copper production whose magnitude can be well appreciated if one considers the construction of the *daibutsu* 大仏 (statue of the Great Buddha) at Tōdai Temple in Nara (*Nara Tōdaiji* 奈良東大寺). Production declined substantially during the 12th to the 14th centuries, but experienced a resurgence at the beginning of the 15th century.

In the early modern era, most Japanese copper was exported, though some was used for the domestic production of copper goods and for coin casting. Exports accounted for more than one-half of production and were always the most significant of the three uses. Indeed, the viability of the copper production industry was dependent on exports. The market for domestically produced goods began primarily with products for the wealthy and powerful, but eventually grew to encompass practical implements of all kinds. Copper coins, a cornerstone of the trimetallic Tokugawa 徳川 coinage system, were also at times in high demand.

This paper describes how Japanese copper production, the three above-mentioned sources of demand and the policies of the Shogunate developed during the Tokugawa period and how they mutually influenced each other.

2 Production Trends

No statistics on copper production or transport are available for early modern Japan. However, documents regarding copper exports over many decades do exist. Additionally, there is some fragmentary data on production and shipment of copper at mines as well as records on shipments, processing and sales in Osaka 大坂, which can be synthesized so as to provide a broad overview.

The period from the end of the 17th to the early 18th century was the golden age of copper in pre-industrial Japan, when production was the highest in the world. Production fell after this, and by the end of the Tokugawa period (1603–1868) it was reduced to one-third its peak level.

 | DOI 10.1163/9789004304512_003

2.1 *Changes in Mine Production Rates*

Copper production began to increase rapidly in the 1660s. For the greater part of the 17th century, the Ashio copper mine (*Ashio dōzan* 足尾銅山) was the largest in Japan. From 1697 to 1699 Japan's copper production reached its peak with an estimated volume of ten million *kin* 斤 (6,000 tons). This is about the same as the production volume at the Besshi copper mine (*Besshi dōzan* 別子銅山) at its peak. During the 18th century the Ani copper mine (*Ani dōzan* 阿仁銅山) was the largest in Japan, followed by the Besshi copper mine. In the mid-18th century, a time when copper production was experiencing a general downward trend, the Ani, Besshi and Osarizawa mines (*Osarizawa dōzan* 尾去沢銅山) were the main sources of copper for export. The Ani and Osarizawa mines were administered by local lords (*ryōshu* 領主). The Besshi mine was managed from the start by the Sumitomo 住友 family.

Because of heavy rainfall and a high water table in Japan, tunnels were prone to flooding, which occasionally caused mining operations to be scaled down or halted altogether. Local lords, city merchants and the local population were keenly interested in the development of copper mines, and again and again mines were redeveloped or newly developed. By the 19th century, however, production had fallen to one-third of peak levels.

2.2 *Primary Mining Regions*

Table 2.1 shows the major mines and their production volumes during the first half of the 18th century, when copper production began to decline.

TABLE 2.1 *Principal mines and their annual production from 1703 to 1721 (Unit: 1,000* kin)[1]

Mine	1703	1708–1715	1716–1721	Province (present-day prefecture)
Osarizawa 尾去沢	300	40–750	0–450	Mutsu 陸奥 (Akita 秋田)
Shirane 白根	60–70	100–330	0	Mutsu 陸奥 (Akita 秋田)

1 Values in the above table for the year 1703 are taken from [Shokoku dōzan], 352–54. Values for 1708–1711 are taken from [Kakitsuke], 3–23. Values for 1712–1715 are taken from [Hikaechō], 45–79. Values for 1716–1721 are taken from [Goyōdō no oboe], 259–89. The figures for the years 1716 to 1721 refer only to copper for export and do not include copper for domestic consumption. Thus, total copper production is actually higher than these figures.

TABLE 2.1 *Principal mines and their annual* (cont.)

Mine	1703	1708–1715	1716–1721	Province (present-day prefecture)
Ateraku 当楽	60–70	0	0	Mutsu 陸奥 (Iwate 岩手)
Mizusawa 水沢	200	0	0	Mutsu 陸奥 (Iwate 岩手)
Kumazawa 熊沢	150–160	0–100	30	Mutsu 陸奥 (Miyagi 宮城)
Shitomae 尿前	100	0	0	Mutsu 陸奥 (Miyagi 宮城)
Ani 阿仁	1,500–1,600	1,400–3,010	1,350–1,700	Dewa 出羽 (Akita 秋田)
Nagamatsu 永松	800–900	160–980	0–280	Dewa 出羽 (Yamagata 山形)
Ashio 足尾	200–300	0	0	Shimotsuke 下野 (Tochigi 栃木)
Kumano 熊野	160–170	40–210	20–70	Kii 紀伊 (Mie 三重 / Wakayama 和歌山)
Tada 多田		50–150	50–100	Settsu 摂津 (Osaka 大坂 / Hyōgo 兵庫)
Ikuno 生野		30–700	300–460	Tajima 但馬 (Hyōgo 兵庫)
Yoshioka 吉岡	60–170	0–60	0	Bitchū 備中 (Okayama 岡山)
Besshi 別子	2,400–2,500	1,050–2,290	850–1,000	Iyo 伊予 (Ehime 愛媛)
Tatsukawa 立川	160–170	0–1,220	410–700	Iyo 伊予 (Ehime 愛媛)
Nobeoka 延岡		10–170	0–70	Hyūga 日向 (Miyazaki 宮崎)
Total production	6,250–6,710	4,300–8,500	3,650–4,610	

2.3 *Life in Mining Towns*

People who lived and worked in mining communities could be roughly divided into two groups: workers (*gezai* 下財) and merchants (*ten'in* 店員). The former included both unmarried workers and married workers whose entire families worked for the mines. The latter were unmarried. Table 2.2 shows the composition of these groups at the Besshi mine.

TABLE 2.2 *Number of workers at the Besshi mine in the years 1713 and 1837*[2]

Type of worker	1713	1837	Description
Diggers (*horiko* 掘子)	450	435	People who dug ore in the mine
Carriers (*ebuhiki* 得歩引)	50	135	People who carried planks and props into the mine and assisted with maintenance
Drainage workers (*mizuhiki* 水引)	150	496	People who drained the mine
Ore-sorting women (*kaname* 砕女)	400	293	The wives and children of those who worked in the mine, who crushed and sorted the ore
Transporters of sorted ore (*hakumochi* 鉑持)	70	106	People who transported ore from the sorting station to the roasting kiln (*yakigama* 焼釜) and from the roasting kiln to the smelting station (*tokoya* 床屋)[3]
Smelter workers (*daiku* 大工)	40	37	People who smelted the ore
Helpers (*tetsudai* 手伝)	200	128	People who operated the bellows and assisted smelter workers
Smiths (*kaji* 鍛冶)	15	13	People who made hammers and chisels for mining

2 [Sumitomo iinkai], *Sumitomo Besshi kōzanshi bekkan*, 78, 101. The overview in Table 2.2 does not include personnel in Tatsukawanakajuku and Niihamakuchiya 新居浜口屋.

3 In early modern Japan furnaces (*toko* 床) were dug directly into the ground. There was no construction on the ground. For details see chapter 2.4 of this paper.

TABLE 2.2 *Number of workers at the Besshi mine* (cont.)

Type of worker	1713	1837	Description
Merchants (*shōnin* 商人) and woodsmen (*kikori* 薪樵)	120	55	Merchants and woodsmen who lived at the mine. Many merchants also commuted from villages.
Day laborers (*hiyō* 日用)	80	87	Workers who constructed bridges and roads in the mine
Woodsmen (*kikori* 木伐)	200	189	People who procured wood for the making of charcoal (the fuel used for smelting), boards and props (mine timber)
Charcoal makers (*sumiyaki* 炭焼)	600	370	People who made charcoal for smelting and for personal use
Charcoal carriers (*suminakamochi* 炭中持)	150	258	People who carried charcoal to the mine
Carriers to Tatsukawa (*Tatsukawa nakamochi* 立川中持)[4]	200	328	People who brought rice and goods for general use from Tatsukawanakajuku to the mine and crude copper from the mine to Tatsukawanakajuku.
Employees of *yamashi* (*yamashi kanai* 山師家内)	100	131	Employees working for the *yamashi* (mining contractors)
Total number of workers	2,825	2,872	
Number of inhabitants in the mining community	4,500	3,348	
Mine output	1,370,000 *kin*	730,000 *kin*	

4 Transport between Tatsukawanakajuku and Niihamakuchiya was by pack-horse (88 horses in 1713, 66 horses in 1837).

The breakdown of workers at the mine by gender was 1,540 men and 463 women in 1721. In 1837, there were 2,262 men and 987 women.[5]

Although the Besshi mine was a large-scale operation, because it was always under the management of a single *yamashi* 山師 (mining contractor), the levels of equipment and personnel were fairly constant. Ashio differed from Besshi in this regard; the mine shafts there were spread over a wide area centered around a single mining town. No one *yamashi* was responsible for the entire mining area. Instead, tens of *yamashi* were responsible for individual mine shafts.

In contrast to those operations in which *yamashi* hired workers and administered a mining town, there were also examples of small-scale mining operations in which local villagers sought investors and managers from among local lords and Osaka merchants. If the mine flourished, the number of workers increased.

2.4 *Copper Refineries in Osaka*

Crude copper that had been processed at a mine was refined and cast into bars for export or into other forms for use in domestic production in Osaka. Two kinds of refining processes were used depending on the type of crude copper. If the crude copper contained more than 0.04375 percent of silver (or more than 26.25 g silver per 100 *kin* of crude copper), silver was extracted using the *nanbanbuki* 南蛮吹 process (also known as the Saiger process). This process consists of three techniques, namely *awasebuki* 合吹 (alloy melting), *shiboribuki* 鉸吹 (liquation, or the method known as *nanbanbuki* in a narrow sense), and *haifuki* 灰吹 (cupellation). Crude copper containing less than 0.04375 percent of silver was processed using *mabuki* 間吹, a technique in which crude copper is heated in a solution to extract impurities. The refined copper was cast into various shapes, for example, into bars for export purposes as shown in figure 2.1.

The *nanbanbuki* process was a method to extract silver making use of lead, which has a stronger affinity to bond with silver than with copper. This process was done on a small scale. The special characteristic of the *nanbanbuki* process is that it required only a small amount of lead. Specifically, the lead was "squeezed out" of the copper by exerting pressure on the alloyed copper using iron tools.

The method called *kobuki* 小吹, whereby molten copper was poured from the crucible into a mold that had been placed in water, was uncommon to Europeans. There are descriptions of the process in the records of visits to copper refineries by the Heads of the Dutch Trading House in Nagasaki. Thus,

---

5 [Sumitomo iinkai], *Sumitomo Besshi kōzanshi bekkan*, 231–32.

FIGURE 2.1 *Casting copper into bars* (saobuki 棹吹). (*in* Kodō zuroku 鼓銅図録 *by Masuda Kō* 増田綱 (*1769–1821*).

COURTESY OF SUMITOMO HISTORICAL ARCHIVES (SUMITOMO SHIRYŌKAN 住友史料館).

for example, the records of Gideon Tant's and Hermanus Menssingh's visits to Sumitomo in 1704 and 1706 respectively describe the refinery there as having 80 furnaces (smelting ovens) in its heyday and give a detailed account of the *kobuki* process.

A copper refinery (*dōfukisho*[6] 銅吹所) was divided into separate spaces for *awasebuki, nanbanbuki, haifuki, mabuki* and *kobuki*. Additionally, there were areas for lead refining (*namarinagashidoko* 鉛流床) where the lead absorbed in the ash, which occurred after the *haifuki* process, was deoxidized to its metallic form. Scenes of this work and pictures of the various tools used can be found in the *Kodō zuroku* 鼓銅図録.

Each furnace was in a small room (*fukidoko* 吹床) separated by walls on three sides. The side facing the corridor was left open. The room was two *ken* 間 (3.6 meters) deep. The width of the room as well as the size and position of the furnace varied slightly depending on the type of furnace. The furnace was made by digging a hole measuring several tens of centimeters in diameter, which was then lined with a mixture of clay, sand and charcoal. The bellows were placed behind a fire protection wall behind the furnace. A hole in the wall allowed the passage of air from the bellows to the furnace. A hood (*hisashi* 庇) was placed on top of the fire protection wall which functioned as a chimney. The furnace was manned by one smelter worker, one bellows operator (two for *mabuki* furnace), and one helper.

Apart from these, the refinery also employed workmen known as *rushi* 留師, who made crucibles, and workmen called *yurimonoshi* 汰りもの師, who collected the copper remaining in the furnaces by using water. Helpers, when they were not working in the furnace rooms, had various jobs such as moving earth, grinding charcoal and preparing for the collection of the copper that remained in the furnaces.[7]

In 1712 the Shogunate ordered seventeen refineries in Osaka to procure and transport copper to Nagasaki 長崎. From then on the number of refineries was fixed. Until that point, when the refineries had been under the control of the First Copper Agency (*Genroku dōza* 元禄銅座, active from 1700 to 1712), there had been no regulation of copper sales and purchases and no restriction on the opening or closing of a refinery. Only the sale of bar copper for export to the Copper Agency had been limited. The seventeen officially designated refineries gave each other mutual support. For example, when a refinery was

---

6 Reading of the author.

7 Sumitomo shūshishitsu, *Kinsei Sumitomo no fukisho no kenkyū*, 17–46; Shōji "Orandajin to Izumiya to Osaka dōfukisho," 71–74.

threatened with closure, it remained in operation with the help of colleagues. Despite such efforts, however, only six refineries remained at the end of the Tokugawa period because of the gradual decline in copper production.

The copper refineries were all located within ca. 1.5 km from the center of the city of Osaka. Each one was positioned on a canal (convenient transport and water supply) and some distance from the city due to its use of fire.

The composition of copper refineries in 1765 is shown in table 2.3. Each entry shows the total number of furnace rooms, minus the number of furnace rooms whose operations were suspended, giving the total number of furnace rooms in operation. Altogether, seventeen refineries existed, but in reality two of them had gone out of business. In addition, one (Kawasakiya 川崎屋, which had no furnace rooms in operation) had stopped operations completely.

TABLE 2.3 *In-service, out-of-service and total number of furnace rooms at refineries in 1765*[8]

Refinery	*awasebuki*	*nanbanbuki*	*haifuki*	*mabuki*	*kobuki*	lead-refining	Total
Sumitomo Kichizaemon 住友吉左衛門	2–1=1	10–4=6	4–1=3	6–0=6	8–3=5	1	31–9=22
Ōsakaya Kyūzaemon 大坂屋久左衛門	3–1=2	10–4=6	4–2=2	3–3=0	6–4=2	0	26–14=12
Ōtsukaya Jin'emon 大塚屋甚右衛門	2–0=2	7–1=6	2–0=2	1–0=1	3–1=2	0	15–2=13
Marukaneya Jirobe'e 丸銅屋治郎兵衛	2–1=1	8–4=4	2–0=2	1–0=1	3–1=2	0	16–6=10
Hiranoya Chūbe'e 平野屋忠兵衛	2–1=1	7–2=5	2–0=2	1–0=1	3–1=2	0	15–4=11
Tadaya Ichirobe'e 多田屋市郎兵衛	2–1=1	8–2=6	2–0=2	1–0=1	3–1=2	0	16–4=12

8 *Dōkata kōyōchō*, 14-6-2.

Refinery	*awasebuki*	*nanbanbuki*	*haifuki*	*mabuki*	*kobuki*	lead-refining	Total
Hiranoya San'emon 平野屋三右衛門	2–0=2	7–2=5	2–0=2	1–0=1	4–2=2	0	16–4=12
Hiranoya Tōemon 平野屋藤右衛門	2–1=1	6–1=5	2–0=2	2–1=1	3–1=2	0	15–4=11
Kumanoya Hikodayū 熊野屋彦太夫	3–1=2	7–2=5	2–0=2	1–0=1	3–1=2	0	16–4=12
Ōsakaya Matajirō 大坂屋又治郎	2–1=1	8–1=7	2–0=2	1–0=1	3–1=2	0	16–3=13
Tomiya Tōkichi 富屋藤吉	2–1=1	6–2=4	2–0=2	1–0=1	3–1=2	0	14–4=10
Tomiya Iemon 富屋伊右衛門	2–1=1	7–3=4	2–0=2	1–0=1	2–0=2	0	14–4=10
Ōsakaya San'emon 大坂屋三右衛門	2–0=2	9–4=5	3–1=2	1–0=1	3–1=2	0	18–6=12
Kawasakiya Katsu 川崎屋かつ	2–2=0	6–6=0	2–2=0	1–1=0	2–2=0	0	13–13=0
Total	30–12=18	106–38=68	33–6=27	22–5=17	49–20=29	1	241–81=160

2.5 *Copper Merchants*

Kyoto 京都 was the seat of the imperial court and since antiquity had been the site of the metal processing industry which manufactured luxury goods. Osaka was a new city, a castle town built as the seat of Toyotomi Hideyoshi's 豊臣秀吉 (1537–1598) government. The city was destroyed by the Shogunate during the Siege of Osaka Castle. When it was rebuilt, it became a center of trade and industry. The metal processing industry in Osaka produced primarily utilitarian goods and raw materials.

Copper exports expanded in scope at the end of 17th century. Japanese refined copper bars of outstanding quality were well received abroad and ordered by the Dutch accordingly. Copper refinery businesses like Sumitomo possessed the technical know-how required to produce such goods. Because Osaka was where copper bars for export were manufactured, it became the centre of the copper industry. Kyoto continued to be the legendary home of luxury goods manufacture, but the greater part of the copper industry was in Osaka.

In Osaka, there were four powerful merchants who not only produced copper bars but were also active in international export as well as the administration of copper mines. Centered around these four merchants were *dōya* 銅屋 who traded copper bars for export, subcontractor refineries, copper merchant houses (*dōtoiya* 銅問屋) who traded crude copper from the mines, copper craftsmen, wholesalers who sold copper in Kyoto and all over Japan, and merchants who dealt in scrap copper. Through the activities of these small merchants specialization in the copper industry increased.

3 Export Trends

In the 16th century, during the so-called Warring States Period (*sengoku jidai* 戦国時代 1467–1573), *daimyō* 大名 (feudal lords) opened new silver and gold mines and increased production of these metals in order to generate funds with which to purchase firearms. The Iwami silver mine (*Iwami ginzan* 石見銀山) is a case in point. At this time, silver from Latin America began to flow first into Europe and then into Asia, bringing the world together in a web of trade. 16th century Japan suddenly became a world leader in the production and export of silver. Japan was a gold-importing nation because the gold to silver ratio there was higher than in other Asian countries. The export of silver and the import of silk yarn and finished fabrics continued in the 17th century, even after the introduction of the so-called Seclusion Policy (*sakoku seisaku* 鎖国政策). In due course, however, silver production declined, and its continuing flow out of the country had a damaging effect. Accordingly, measures were taken to check the outflow of silver by increasing copper production and making copper Japan's principal export product. And copper, by replacing silver as principal export product, helped finance the import of goods such as silk yarn and finished fabrics as well as medicine. From the 18th century onward, domestic production of these import goods continued to grow, thus reducing the country's dependence on imports.

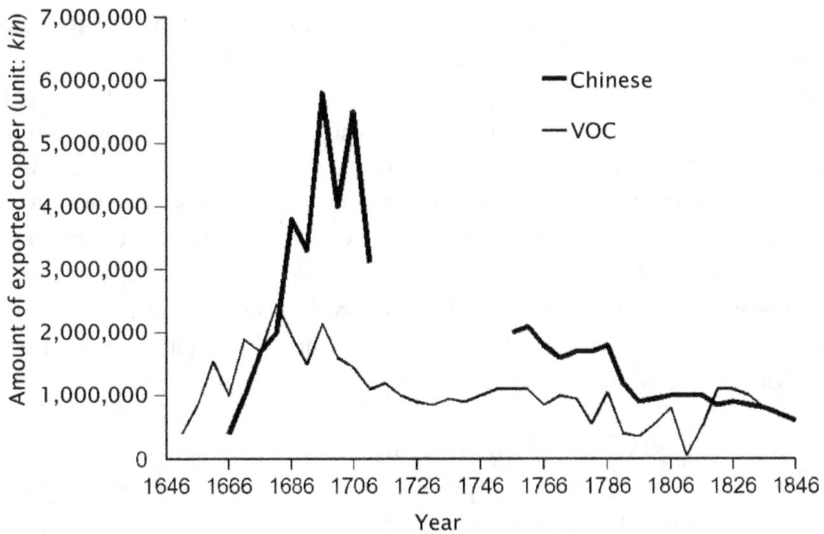

FIGURE 2.2 *Volume of copper exports from Nagasaki 1646–1846.*[9]

3.1 *Changes in Export Quantities*

The changes in the volume of copper exported from Nagasaki are illustrated in figure 2.2. Comparing the quantities carried by Dutch and Chinese ships, the Dutch transported the greater amount in the mid-17th century, but this trend was reversed in 1682. In 1685, the Shogunate implemented the trade law known as *jōdakasei*[10] 定高制 which restricted trade value to a maximum of 3,000 *kanme* 貫目 in silver currency to the Dutch and 6,000 *kanme* to the Chinese. This *jōdakasei* did not regulate the export of copper, but the *Shōtoku shinrei* (正徳新例 Shōtoku New Law) enacted in 1715 set a ceiling on the volume of copper exports, allowing only three million *kin* to be shipped on Chinese vessels and 1.5 million *kin* to be shipped on Dutch vessels. These limits were frequently changed, but the ratio was always maintained. Actual export rates were also dependent on circumstances.

3.2 *The Copper Trade at the Dutch East India Company*

The Dutch East India Company (VOC: *Vereenigde Oostindische Compagnie*), unlike the Portuguese who had been the first Europeans to arrive in Japan, bent to the will of the Shogunate and engaged only in trade rather than attempting to convert the locals to Catholicism. Its trading house in Japan (first in Hirado

9 Shimada, "Kinsei Nihon no dōyushutsu sakugen to Chōsen no senkō," 30.

10 Reading of the author.

平戸 and later in Nagasaki) was consistently the most profitable of all its trading bases.

The VOC sold much of the copper that it brought from Japan in India. The remainder was sold to Chinese ships in Batavia (Jakarta). The copper sold in India was traded for Indian goods that were in turn traded in Southeast Asia for goods to be sold in Europe and Japan. Thus, copper was an important product that allowed the VOC to generate profit from intra-Asian trade. Japanese copper was generally sold in Asia, though it was also sold in Europe in times of war. Because copper bars became the standard of the international export market for copper, Osaka, which had mastered the art of their manufacture, became the center of the copper market in Japan.

3.3 *Replacing Silver as Principal Export*

Even after the implementation of Japan's *sakoku* policy, silver remained a principal export product. From 1648 to 1667, the quantity of silver coins, Keichō (*Keichō gin* 慶長銀), exported on Chinese and Dutch ships was on average approximately 14,000 *kanme* (52.5 tons) per year.[11] The Keichō silver coin was 80 percent pure, which means that 42 tons of pure silver were exported annually. Such large outflows of silver were not propitious at a time when the production of silver mines was declining and currency was becoming increasingly important for domestic economic development. The Shogunate responded with various regulations restricting the export of silver. In 1685, when the *jōdakasei* was implemented, copper, with its growing production levels, replaced silver as a principal export good.

3.4 *Trade with Chinese Ships*

During the golden age of Japanese copper from the end of the 17th century through the early 18th century, China used Japanese copper as an ingredient in its currency during a lull in domestic copper production that occurred between periods of high production in the Song Dynasty (960–1279) and the Qianlong period (1736–1795) of the Qing Dynasty. Until the beginning of large-scale copper production in Yunnan Province in the mid-18th century, Chinese ships purchased massive quantities of copper from Japan. As indicated earlier, they also purchased some of the copper that had been exported from Japan by the VOC in Batavia.

11 Sumitomo shūshishitsu, *Daiichiji dōza to Sumitomo*, 21.

4 Trends in Domestic Processing and Consumption

Originally, copper was used for roof tiles for prestigious buildings, statues of Buddha, temple bells, lanterns, decorations for bridge balustrades and luxury products in response to demand from wealthy and powerful individuals. In time the use of copper increased, e.g., for *funatekazari* 船手錺 (a type of decorative hull reinforcement on Japanese ships) and items for daily use.

4.1 *Forms of Copper*

Copper was cast into different forms, as listed in table 2.4. The *Kodō zuroku* contains illustrations of the molds used for casting copper as shown in figure 2.3.

TABLE 2.4 *Various forms of copper for domestic consumption and their uses*[12]

Japanese	English	Use
teidōrui 丁銅類	Various types of plates	*funatekazari*, roofing tiles, fire shovels, small tools
ōteidō 大丁銅	Large plates	large *funatekazari*
toiteidō 樋丁銅	Gutter plates	various kinds of pipes and water conduits
denchūdō 伝中銅	*denchū* copper	medium and large pipes and water conduits
naganobedō 長延銅	Oblong ingots	all kinds of *funatekazari*
nagasaodō 長棹銅	Long bars	wire
shakushidō 杓子銅	Ladle copper	ladles, roofing nails, pegs and tacks
chūhiradō 中平銅	Medium-sized flat copper	roofing nails, pegs and tacks
ōmarudō 大丸銅	Large discs	boilers for bathing, bathtubs, braziers
komarudō 小丸銅	Small discs	Tea kettles, medicine containers, plates, oil cans, cooking pots, water bath pots, various kinds of small round implements
Shiboridō 鉸銅 and *mabukidō* 間吹銅	Copper refined using *nanbanbuki* or *mabuki*	For the production of alloys such as brass and bronze, not for casting into forms

12 *Dōkata kōyōchō niban*, 14-6-1.

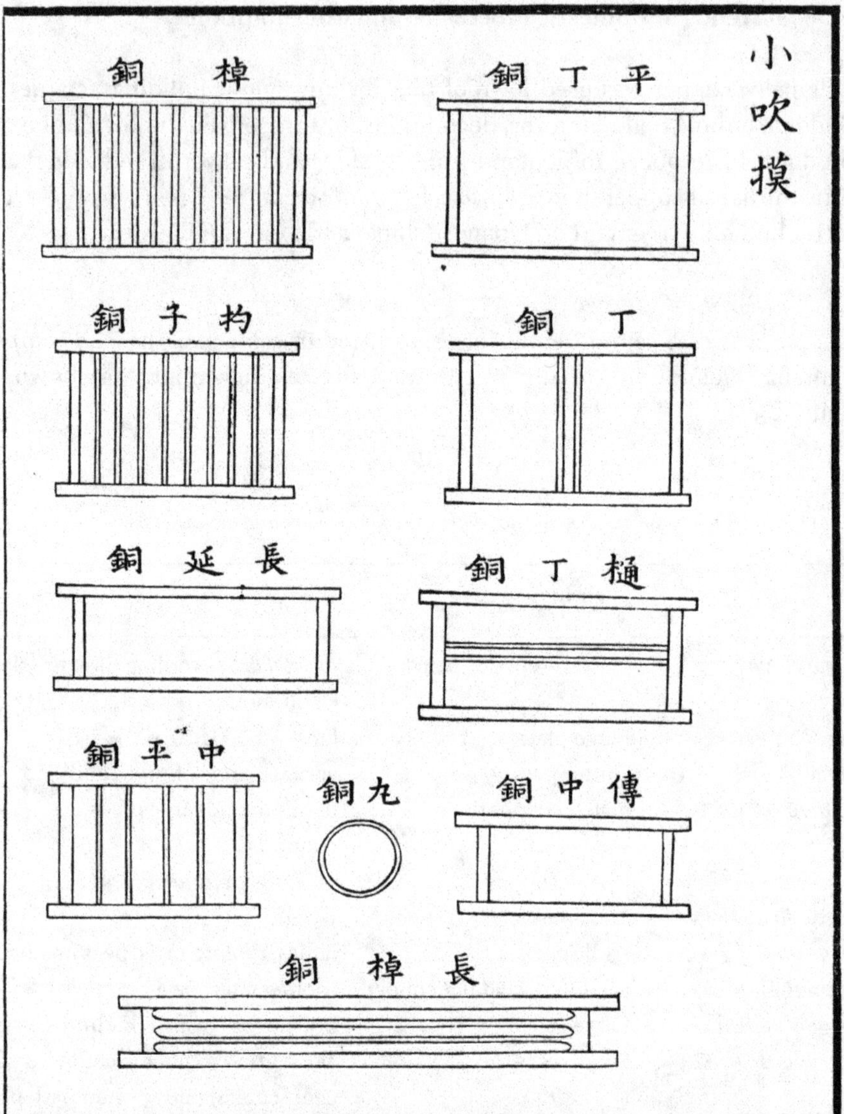

FIGURE 2.3 *Molds for casting* (*in* Kodō zuroku 鼓銅図録 *by Masuda Kō* 増田綱 (*1769–1821*).
COURTESY OF SUMITOMO HISTORICAL ARCHIVES (SUMITOMO SHIRYŌKAN 住友史料館).

4.2 *The Development of the Brass Industry*

Brass is an alloy composed primarily of copper and zinc that has a great many uses. Because the color of brass can be varied by altering its composition, there was increasing demand for imitation gold, silver and copper as well as brass products for items for everyday use. Initially, brass was produced in Kyoto and later in Osaka, Sakai 堺, Fushimi 伏見 and Edo 江戸 (present-day Tokyo). During the Edo period (1603–1868) zinc, which was required for the manufacture of brass, could not be produced in Japan and had to be imported. The amount of imported zinc fluctuated quite considerably, with the highest annual import volume reaching 400,000 to 500,000 *kin*. Apart from being used to make alloys, zinc was used to coat iron plate and for roofing. Brass contained up to 30 percent zinc and twice or four times as much copper. In the latter half of the 18th century, as copper production began to decline, scrap copper (*kodō* 古銅) came to be used more commonly in brass manufacture.

5 Trends in Coin Production

A currency-based economy began to develop in medieval Japan during the Muromachi period (*Muromachi jidai* 室町時代 1336–1573), when copper coins imported from China, as well as imitations of these, were in circulation. As Japanese production of silver and gold increased in the 16th century, higher-denomination currency in these metals began to circulate as well. Gold coins were most widely used in eastern Japan, where the country's gold was predominantly mined, and silver coins were more often in circulation in the west, where many of the silver mines were located. This development led to the introduction by the Shogunate of a trimetallic coinage system using gold, silver and copper coins.

5.1 *The Trimetallic Currency System*

The trimetallic coinage system introduced by the Shogunate was known as the *sanka seido* 三貨制度 (literally "three coin system"). The official rate of exchange established in 1609 was one *ryō* 両 of gold (one *koban* 小判 coin) to fifty silver *chōgin* 丁銀 coins or 4,000 copper *zeni* 銭 coins.[13] In 1700, this was changed to one *ryō* gold to sixty *me* of silver, though the relative value of gold and copper remaining unchanged. The relationship between the three

---

13 Commonly the unit *kanmon* 貫文 was used to count a large amount of copper coins. One string consists of 1,000 copper coins. 4,000 copper coins are equivalent to 4 *kanmon zeni* coins.

currencies was not one of primary and secondary currencies, but of three separate currencies whose exchange rates the Shogunate attempted to keep consistent with both the market value of the metals they contained and official standards.

After the trimetallic currency system had been established, the Shogunate introduced the policy of debasing gold and silver currency and using the resulting seigniorage to finance its budget deficit. This policy served the dual purpose of providing the government with a source of revenue and of increasing the money supply as the economy developed. It was continued for many years.

As money supply grew, it became necessary to cast an increasing amount of copper coins. A large number of copper coins were cast in the years 1637–1640 and 1668–1683, and even greater quantities were produced during 1700–1706 and 1736–1745. In this final period from 1736 to 1745, some 6,760,000 *kanmon* were cast, a quantity requiring an estimated 23,820,000–30,620,000 *kin* (ca. 14,000–18,000 tons) of copper.[14] This is almost equivalent to the volume of exported copper, which normally accounted for the lion's share of copper consumption.

Though the trimetallic currency system officially remained consistent throughout the Edo period, in fact the Shogunate continued to issue silver currency with the value of gold *ryō*, and to gradually reduce the purity of gold and silver coins. The result was that silver ultimately became a secondary currency, and that by the fall of the Shogunate Japan had effectively adopted the gold standard.

5.2 *The End of the Production of* Kan'ei tsūhō *Copper Coins*

In the middle of the 18th century, the price of copper rose to the extent that the cost of casting copper coins was more than they were worth. As a result the production of *Kan'ei tsūhō dō ichimonsen* 寛永通宝銅一文銭 (one-*mon* copper coins) was stopped: One of these copper coins contained one monme (3.75 g) of copper. This means that 4,000 coins contained 25 *kin* of copper. Since the official value ratio of silver coin to copper coin was 60 *me* to 4,000 *mon*, there would be no profit when 25 *kin* of copper were worth more than 60 *me* silver. The price of copper rose to about this level (e.g. 100 *kin* copper cost 240 *me* silver) in the period 1736 to 1745, during which large quantities of copper *ichimonsen* were cast. Thus the copper coin *Kan'ei tsūhō ichimonsen* could no longer be viably produced. From this time on, iron *ichimonsen* as well as four-*mon* brass coins were produced as *Kan'ei tsūhō*. By the end of the Edo period, hundred-*mon* (*Tenpō tsūho* 天保通宝) and four-*mon* (*Bunkyū eihō*

14 Nihon ginkō chōsakyoku, *Zuroku Nihon no kahei*, 147; Kobata, *Nihon dōkōgyōshi no kenkyū*, 731.

文久永宝) copper coins were also being minted whose face values were much higher than the actual value of their copper content.

6 Market Trends and Control by the Shogunate

In the 17th century the domestic copper trade was basically a free market. The Shogunate occasionally banned copper exports during periods when copper coins were produced on a large scale, but at other times anyone was free to export copper in whatever quantities the market demanded, with prices negotiated by the traders.

Starting in the 1670s, the Shogunate began to exert more control over the export of silver. In addition, in 1684, China lifted a ban on the international movement of its trading ships. Starting the following year, the number of Chinese ships arriving in Japan rose drastically, and without intervention would have caused a correspondingly large increase in the outflow of silver. In 1685, the Shogunate responded by setting a trade ceiling of 3,000 *kanme* in silver currency on Dutch ships and 6,000 *kanme* on Chinese ships. These limits were called the *jōdaka* 定高, and the law that established them was called the *jōdakasei*.

The regulation was simple: By enforcing the above ceiling levels and turning Chinese ships back without allowing them to do trade if the limits had been exceeded, it was possible to halt the increase in silver outflow despite the rise in Chinese merchant traffic.

As for the copper trade, it was largely unchanged up to that point; the price was established by negotiation, and the quantities exported were determined by market forces. Copper exporters were limited to the traditional sixteen copper merchants. As copper production grew and the outflow of silver was brought under control, it became possible to establish copper as a principal export good, and the role of copper became increasingly important.

6.1 *The International Copper Trade in the 17th Century*

During the 17th century, copper production grew and trade was unregulated. Table 2.5 shows copper exports during the two years prior to the establishment of the *jōdakasei* and the two years following.

The enforcement of the *jōdakasei* in 1685 led to an increase in the number of Chinese ships arriving at Japanese ports and in the quantity of copper exported. As indicated in the table, some of the ships were turned back without being permitted to trade their goods.

TABLE 2.5 *Numbers of Dutch and Chinese ships arriving in Japan and the quantities of copper exported, 1683–1686*[15]

Year	Chinese ships (+ ships sent back)	Dutch ships	Copper exported on Chinese ships (in 1,000s of *kin*)	Copper exported on Dutch ships (in 1,000s of *kin*)	Total (in 1,000s of *kin*)
1683	27	3	2,330	1,600	3,930
1684	24	5	2,610	2,280	4,890
1685	73 (+ 12)	4	2,330	2,100	4,430
1686	84 (+ 18)	4	3,240	2,000	5,240

Copper production and the number of Chinese ships arriving in Japanese ports continued to increase from this time on. A build-up in copper inventories due to Chinese ships being turned away, fears of smuggling by the Chinese ships denied the right to trade, and the increased demand for the goods carried by Chinese ships all brought pressure to bear on the *jōdakasei*, especially as regards the limitations on trade with the Chinese. Because of this, a system known as *shiromonogae* 代物替 was put in place, by which the Shogunate allowed additional copper to be traded in exchange for a portion of the profits generated. Table 2.6 illustrates the development of copper exports following the introduction of the *shiromonogae*.

The quantities of copper exported in the years 1696–1698 amounted to almost nine million *kin*, as can be seen in table 2.6. This is because stockpiles of copper were being exported. It should not be assumed that such a high level of export was to be expected each year.

The total copper exports indicated in table 2.5 are the quantities permitted by the *jōdakasei*. This quantity is indicated in the column "*jōdaka*" in table 2.6. A comparison of tables 2.5 and 2.6 makes clear that the quantities permitted by the *jōdakasei* decreased. The growth in copper exports during the period covered in table 2.6 is due to the introduction of the *shiromonogae*.

15 *Nagasaki jitsuroku taisei*, 220–1, 262; *Nagasaki jikki nendairoku*, 124–60; [Kaiwatashi dōdakachō].

TABLE 2.6 *Numbers of Chinese and Dutch ships and quantities of copper exported (per* jōdakasei *and* shiromonogae*), 1696–1699*[16]

Year	Chinese ships (+ ships sent back)	Dutch ships (+ ships sent back)	Copper exported on Chinese ships (in 1,000s of *kin*)	Copper exported on Dutch ships (in 1,000s of *kin*)	Total (in 1,000s of *kin*)	*jōdaka* ceiling (in 1,000s of *kin*)	*shiromonogae* (in 1,000s of *kin*)
1696	70 (+ 11)	4	6,930	1,830	8,760	3,720	5,040
1697	70 (+ 33)	6	6,400	2,500	8,900	3,940	4,960
1698	68 (+ 3)[17]	6 (+ 1)	5,900	2,500	8,400	3,770	4,630
1699	69 (+ 4)	5	4,970	2,500	7,470	2,970	4,500

6.2 *The Copper Trade of the 18th Century*

From the period described in table 2.6 onward, copper exports experienced a series of rapid fluctuations. The first consequence of these fluctuations was the establishment of the First Copper Agency, the *Genroku dōza*, in 1701. The fluctuations continued, however, until the establishment of the Third Copper Agency, the *Meiwa dōza* 明和銅座 (Meiwa Copper Agency) in 1766.

The *Genroku dōza* was positioned above the copper refineries, from which it purchased bar copper that it then exported. The Agency forced twelve of the sixteen traditional copper merchants (*dōya*) out of business, leaving only four who operated copper refineries in addition to their copper trading businesses. The *Genroku dōza* purchased copper from refineries for less than the market price. Even if the market price of copper increased, the export price was kept stable, and the refineries were forced to accept lower sales prices and a loss of revenue. Due to financial difficulties at the Copper Agency, the transport of bar copper to Nagasaki was taken over by an association of refineries (*dōfukiya nakama* 銅吹屋仲間) in 1712–1715.

In 1715, the Shogunate introduced regulations on imports and exports passing through Nagasaki, known as the *Shōtoku shinrei*. These detailed and comprehensive regulations re-established limits for exports that had gradually increased to more than their original levels following the introduction of the

16 Ibid.

17 These three ships were not sent back, but their cargo was lost due to fire.

TABLE 2.7 *Numbers of Chinese and Dutch ships and quantities of copper exported (per* jōdakasei *and* shiromonogae*), 1711–1714*[18]

Year	Chinese ships (+ ships sent back)	Dutch ships	Copper exported on Chinese ships (in 1,000s of *kin*)	Copper exported on Dutch ships (in 1,000s of *kin*)	Total (in 1,000s of *kin*)	*jōdaka* (in 1,000s of *kin*)	*shiromonogae* (in 1,000s of *kin*)
1711	57	4	4,190	1,000	5,190	1,940	3,250
1712	59 (+ 3)	4	2,710	750	3,460	1,560	1,900
1713	40 (+ 9)	3	3,580	1,000	4,580	1,930	2,560
1714	51 (+ 1)[19]	3	3,850	1,050	4,900	2,310	2,590

jōdakasei. As in the past, the trade ceiling was set at 6,000 *kanme* in silver currency for Chinese ships and 3,000 *kanme* for Dutch ships. In addition, a new rule was introduced limiting copper exports to 3 million and 1.5 million *kin* for Chinese and Dutch ships respectively. The status of copper exports immediately prior to the introduction of the *Shōtoku shinrei* is illustrated in table 2.7. Compared to the period detailed in table 2.6, export quantities are lower; the new regulations were in line with the level of copper production which had also begun to decline. The decrease should thus not be seen as evidence that the Shogunate had succeeded in limiting exports, but rather that the decline in copper production limited the amount of copper available for export.

As the price of copper rose in 1738, the Shogunate established the Second Copper Agency (the *Genbun dōza* 元文銅座), which monopolized the copper market. The Agency bought up the entire supply of crude copper at prices reflecting the state of the market, had it processed by copper refineries, and then either exported it via Nagasaki or sold it for domestic manufacturing. The price of copper for the Chinese and Dutch remained low, and losses were compensated by the profits earned from the domestic sale of imported goods. This became the standard model of copper exports from this point forward. When the market price of copper fell soon after, the price of copper for export

18 *Nagasaki jitsuroku taisei*, 220–1, 262; *Nagasaki jikki nendairoku*, 124–60; [Kaiwatashi dōdakachō].

19 This ship was wrecked off Hirado.

(copper from the Ani, Besshi and Osarizawa mines), which the mines sold to the Copper Agency, was brought down to this new market average, where it was then fixed.

Copper prices rose again in 1765. However, because the price of copper for export was fixed, this increase was only in the price of copper sold on the domestic market. As mentioned above, the Shogunate then established the *Meiwa dōza*, the Third Copper Agency, which again had a monopoly on copper, buying it at a price established by adding refinery costs as well as commission for copper merchants and the Copper Agency to the production cost at the mine. This stabilized the domestic copper market. In the following years, there were occasional fluctuations, but the foundation laid by the *Meiwa dōza* left the copper trade largely stable.

7 Conclusion

Japanese copper production began to grow rapidly in the early 1660s, and had become the largest in the world by the end of the century, with approximately ten million *kin* (6,000 tons) annually. Production declined after this and was only around one-third of its peak level by the end of the Edo period. Copper was mainly exported throughout the early modern period and replaced silver as Japan's principal export at the end of the 17th century. It was occasionally in huge demand for coin production, but due to rising copper prices, it became unviable to cast *dō ichimonsen* (one-*mon* copper coins). However, the production and consumption of copper items for everyday use increased.

On the one hand, trade ceilings and laws regulating Nagasaki trade varied according to developments in production and domestic trade, and the significance of Nagasaki trade began to decline in the 18th century. On the other hand, copper production and domestic trade were variously influenced by the need to maintain international trade, but if one considers the bigger picture, the domestic production of goods that were dependent on imports was able in time to become less reliant on international trade. It was no longer necessary to export copper in order to import these goods. Accordingly, domestic production and consumption of copper grew.

CHAPTER 3

The Akita Domain and Osaka Merchant Houses at the Time of the Establishment of the Meiwa Copper Agency

Iwasaki Yoshinori 岩崎義則

1 Introduction

It is generally understood that copper production in early modern Japan was severely restrained by the Tokugawa Shogunate's monopoly and low-price policy, as suggested among others by Ishii Takashi 石井孝.[1] However, when taking into account the involvement of Osaka merchant houses in the copper business, a new picture emerges.[2] In the Akita domain (*Akita han* 秋田藩), the Osaka copper merchant house (*Osaka ton'ya* 大坂問屋) Nagahamaya 長浜屋 financed the copper mine and took the responsibility for copper transport to Osaka; it generated sufficient income to compensate for the high costs of mine management and to cover the domain's expenditure in Edo. Thus, not only the Shogunate's policy on copper regulation, but also the Osaka copper merchant houses significantly influenced copper production. I will demonstrate this by

---

1 See Ishii, "Bakumatsu ni okeru bakufu no dō yushutsu kinshi seisaku." A recent study by Imai Noriko 今井典子 is an exception. Based on a comprehensive study of copper production in the Edo period, she proposed a new interpretation of the function of the Shogunate's copper monopoly, that is, the monopoly was created not for the purpose of restricting copper production but in response to the maturation of the copper market. Imai supported this argument by regarding copper in early modern Japan as a commodity (Imai, "Kinsei Nihon no dō: dōishijō to dōtōsei"). My study concurs with this new approach to the early modern copper business by delineating the Akita domain's relationship with the Nagahamaya merchant house and demonstrating that the influence of the Shogunate's copper monopoly on the domain's copper production was limited.

2 Some studies on the involvement of Osaka merchant houses in the copper business have been published, though there are still desiderata. For instance, Sasaki Junnosuke 佐々木潤之介 closely examines the Akita domain's operation of its copper mine from the perspective of mine history. Regarding the question how the income from the sales of copper for overseas export was invested for the domain's copper production, however, he only analyzed the amount of silver payment for copper but not the role of Osaka merchant houses. See Sasaki, "Kinsei sandō seisaku ni kansuru ichikōsatsu," 11.

 | DOI 10.1163/9789004304512_004

scrutinizing the relationship of the Akita domain with Osaka merchant houses around the time of the establishment of the Meiwa Copper Agency (*Meiwa dōza* 明和銅座 or the Third Copper Agency) as evinced by the diaries of Ishii Shōzaemon Tadayuki 石井庄左衛門忠運, who was a *motokata bugyō* (financial magistrate 本方奉行) of the Akita domain.[3]

2 The Akita Domain and the Osaka Copper Merchant House Nagahamaya

2.1 *Shipment of Akita Copper to Osaka*

Table 3.1 shows the sales of Akita copper from the end of the Hōreki 宝暦 period (1751–1763) to 1774. Until 1762, the domain sold the annual quota of 1,800,000 *kin* 斤 (1,080 tons) of copper. However, the volume of copper supplied to Osaka subsequently decreased. As a result the amount of copper reaching Nagasaki 長崎 stagnated and this in turn caused a serious decline in copper exports to foreign countries. In 1763 sales fell below the pre-determined quota and in 1764 and 1765 amounted to only about half the level of 1762. From 1766, there was a sign of recovery and relatively stable sales continued until 1774. How can the decline and recovery of Akita copper sales before the establishment of the Third Copper Agency in 1766 be explained?

While the fluctuation in sales of Akita copper was quite an important issue, no studies have hitherto attempted to find the reason for this development. In order to do so, I analyze copper mine management and the domain's finances which were closely linked to the fluctuation in copper sales.

Let us briefly outline the relationship between the Akita domain and Osaka copper merchant houses prior to 1761. The Ani copper mine (*Ani dōzan* 阿仁銅山) opened in Akita during the Kanbun 寛文 period (1661–1672) and the domain took over the direct management of the mine in 1701. In 1708 the total production of copper and lead amounted to 36,000 *ko* 箇 which was equivalent to 3,600,000 *kin*. As a *tedai* 手代 (clerk) of the mine wrote, Akita's copper

3 *Ishii Tadayuki nikki*. The diary covers the years from 1757/7 to 1790/7, or from the age of thirty-eight to the age of seventy-one. The diarist's records during his term as a magistrate of finance are particularly detailed. Together with *Kokuten ruishō* the diary is a significant source for the study of the domain's government during the Edo period. The translation into modern Japanese was published in May 1973 as part of the second series of *Shin Akita sōsho dainiki* 新秋田叢書第二期 but I consulted the original text as much as possible. As for the quotations in this essay, I have mentioned the dates of the diary entries rather than the page numbers of *Shin Akita sōsho dainiki*.

TABLE 3.1 *Sales of Akita copper (unit:* kin, *100* kin = *60 kg)*[4]

Year	AD	Copper for overseas export	Copper for domestic consumption	Total
Hōreki 10	1760	1,800,000		1,800,000
11	1761	1,800,000		1,800,000
12	1762	1,800,000		1,800,000
13	1763	1,422,558		1,422,558
Meiwa 1	1764	1,031,994		1,031,994
2	1765	1,036,500		1,036,500
3	1766	1,036,500	505,809	1,542,309
4	1767	1,036,500	388,794	1,425,294
5	1768	1,036,500	603,346	1,639,846
6	1769	1,036,500	406,046	1,442,546
7	1770	1,236,500	402,357	1,638,857
8	1771	1,461,500	45,507	1,507,007
An'ei 1	1772	1,461,500	303,223	1,764,723
2	1773	1,461,500	37,613	1,499,113
3	1774	1,461,500	173,975	1,635,475

production was indeed entering a period of prosperity and making extensive contributions to the domain's finances.[5]

How was Akita copper sent to Osaka? Below is a brief description of the situation at that time:

> After the copper mine was placed under the direct management of the domain, copper was transported by *yamashi* 山師 (mine brokers) from Akita to Otsu 大津 and was sold at the prevailing market price. Meanwhile Osaka merchant houses were put in charge of copper. They put the copper for sale and sold it to copper refineries and others in Osaka.[6]

4 *Daiisho*, vols. 3 and 4.

5 "The copper mines and their success can be considered to have reached their zenith." 此節之御山之盛り昼時と可申哉ニ候. "Akita-gun Ani-dōzan shidai kikigaki."

6 其（銅山直営）以来は年々之出銅御山師共え被仰付、大津江被差廻御直段次第御払被成置、其後大坂問屋共江支配被仰付、大坂惣吹屋共入札ニ而御払被成置、御山仕入は勿論、自然ニ御財用御融通ニも相成候. *Kokuten ruishō*, 425.

This indicates that Osaka copper merchant houses took over the transport and sale of Akita copper from the *yamashi*.

In the midst of the Shogunate's monopoly over copper production and the Akita domain's financial crisis, Nagahamaya Genzaemon 長浜屋源左衛門, the owner of the Osaka merchant house Nagahamaya, gained power and influence over copper production in Akita.[7] It is documented that, in 1726, Nagahamaya Genzaemon, Yamashita Hachirōemon 山下八郎右衛門, and Hiranoya Matabei 平野屋又兵衛 served in the copper merchant houses for the Akita domain.[8] The domain, however, became solely dependent on Nagahamaya after Hiranoya resigned in 1739[9] and Yamashita and Takaoka Kichiemon 高岡吉右衛門, who were also *kuramoto* 蔵元 (merchants who handled rice and special products from the domains in the Edo period), were dismissed.[10] Thereafter, the domain's dependence on Nagahamaya increased further as the financial strength of the influential copper refinery Osakaya 大坂屋 deteriorated, which cooperated with the domain during the time when the domain directly managed the mine: In 1772 Osakaya had to ask the domain to lend it 100 *ko* of copper and 10 *kanme* 貫目 of silver money, and in the following year it was forced into temporary closure.[11]

A significant characteristic of Akita's finances was that Nagahamaya was also a *kuramoto* and hence also dealt with rice. From the following oral message in 1768/8, we know that the Akita domain paid off the debt to Nagahamaya with rice, which was not only food, but also functioned as a currency at that time:

> We honour our debt to Nagahamaya with next year's rice shipment to Osaka. We request silver investment with next year's rice shipment to

---

7 With regard to Nagahamaya's lineage, the following article from 1771/9/2 (Meiwa 8) in *Ishii Tadayuki nikki* can be referred to: "Around the time from Kan'ei 14 to 15 (1637 to 1638), there was one letter on a stone lantern donated by Nagahamaya Genzaemon to the Sumiyoshi Shrine 住吉大社 that I could not decipher. Thus, Genzaemon wrote his name again at Itamiya and had a stone mason inscribe it. This was about 135 years ago from this year, which is 1771." 寛永十四五年住吉へ長浜屋寄進の石灯篭一文字見得兼候付伊丹屋にて源左衛門新に銘を書、石工に為彫候、当夘年迄百三十五年の由. *Ishii Tadayuki nikki* 1771/9/2 (Meiwa 8).

8 *Dōdaigin*.

9 Akitaken, *Akitaken shi shiryōhen*, 344.

10 潰. *Ishii Tadayuki nikki*, 1769/4/4 (Meiwa 6).

11 Sasaki, "Osaka dōton'ya Osakaya ni tsuite no oboegaki," 1–18.

> Osaka as collateral. The amount of rice we will pay next year depends on your silver investment.[12]

The fact that the copper transport route was changed from the route via Otsu and Lake Biwa to the so-called westbound marine route, which directly connects present-day Hokkaidō to Osaka, explains why the Osaka merchant house Nagahamaya gained increasing influence on the copper business of the Akita domain. And because rice was also transported to Osaka, the ties between Nagahamaya and the domain were further strengthened.[13] In order to make a thorough investigation of the stagnation of the domain's copper shipments to Osaka, which started in 1763, it is essential to consider Nagahamaya's involvement in both the copper and the rice business.

2.2 *The Expulsion of Nagahamaya*

On 1761/7/24, the following letter reached Akitaomote 秋田表 from Ishizuka Ichinokami 石塚市正, a *karō* 家老 (senior retainer) who was visiting Osaka:

> This time Ichinokami was commanded, together with *motokata bugyō* Tan Sōjūrō 丹宗十郎, to travel to Kyoto 京都. I was not informed of the details. I heard that Ichinokami and Tan travelled because neither in Akita nor in Kamigata 上方 (Kyoto-Osaka area) were there the financial resources to secure payment to Edo 江戸 before the season of *obon* お盆.[14] In this situation, on the third day of this month, I was informed that all the *kuramoto* and copper merchant houses in Kamigata had been expelled.[15]

---

12 昨今出銀之分、来御廻米にて引取候而は不相成間敷候間、来廻米に向、唯今より出銀相頼候、其受高次第来廻米相渡旨申談候. *Ishii Tadayuki nikki*, 1768/8/28 (Meiwa 5).

13 What is important especially in this case is that, triggered by the establishment of the westbound route, Osaka took over from Kyoto as the country's commercial centre. In 1680, the Akita domain owed 4,940 *kanme* of silver to Edo, Kyoto, and Akita but was not yet in debt to Osaka. In 1690, however, Takaoka Shigemasa 高岡重政 was charged with managing the debt owed to Osaka. This shows that the tie between the domain and the capital in Osaka had become stronger (*Umezu Tadayoshi nikki*). Incidentally, the relationship between the above-mentioned Takaoka and Takaoka Kichizaemon 高岡吉左衛門, whom I will mention later, cannot be clarified.

14 *Obon* is the Japanese festival of the dead in the middle of the summer.

15 此度市正殿上京被仰付、丹宗十郎（元方奉行）同道罷登候、此方へは相談も無之具の儀不相知候得共、当盆前江戸御仕送、秋田・上方よりの手当無之に付、右御用被仰付上京の由、左候得は、当三日上方表御蔵元不残御用

The letter concerns the expulsion on 1761/7/3 of Nagahamaya who, as indicated earlier, was both a copper merchant and *kuramoto*. Previous studies have interpreted this expulsion simply as a measure resulting from the domain's financial crisis and did not proceed to examine its remarkable aftermath.[16] After the expulsion, the financial crisis actually escalated to the extent that the Akita domain's finance department ironically tried to reinstate Nagahamaya.

While Ishii, a finance magistrate of the Akita domain at the time, mentions in his diary that "[he does] not know the details"[17] about the reason for Nagahamaya's expulsion, the fall in the price of Akita rice can be considered one of the major causes. Around 1761 the price of Akita rice fell and in winter was down to 31 *monme* 匁 and 5 *bu* 分 of silver per *koku* 石 (ca. 180 litres, about the annual consumption of an adult Japanese), although winter was normally the season when the price of Akita rice became relatively high.[18] This price erosion led to a decline in the collateral value of rice. The Akita domain presumably attempted to resolve this problem by expelling Nagahamaya and thus reneging on its obligation to him, namely the payment for the debt amounting to 300,000 *ryō* 両 of silver [sic].[19] The fact that the copper mine's financial crisis escalated after the expulsion of Nagahamaya suggests that the domain's rice shipments to Osaka significantly influenced its copper shipments and sales.

According to Ishii's diary, although there was a shortage of about 1754 *kanme* of silver money at the copper mine in 1762, funds were sufficient for that year. It was reported that a further shortage of 258 *kanme* of silver for the coming year was inevitable.[20] Indeed from 1763 on, the lack of funds started to seriously affect the management of the copper mine.

After 1763/3, there was no prospect of the silver supply from Kamigata going to the copper mine and the financial situation worsened to the extent described in the following diary entry:

> Because silver money has not been transferred from Kamigata to Akita, we cannot allocate capital as planned. [...] Due to this situation [...]

召放、銅問屋も別人可被仰付にて御用被召放候段申来候. *Ishii Tadayuki nikki*, 1761/7/24 (Hōreki 11).

16 See Sasaki, "Kinsei sandō seisaku ni kansuru ichikōsatsu."

17 具の儀不相知.

18 See the table of rice prices in Sasaki, *Dōjimamai shijō-shi*.

19 *Kokuten ruishō*, 13. Here 300,000 *ryō* of silver means silver that has a value equivalent to 300,000 *ryō* of gold.

20 *Ishii Tadayuki nikki*, 1762/5/11 (Hōreki 12).

> from this spring, there will be no money to provide for the copper mine. Unless money is supplied now, there will be a riot among mine workers.[21]

The weak financial capability of Saikaya Shichibei 雑賀屋七兵衛, a new copper merchant house on whom the domain had to rely after expelling Nagahamaya, restricted the supply of silver money to the Akita copper mine and thus aggravated the financial crisis there: "if silver money transfer to the copper mine is delayed by one day, the loss will be enormous. From now on, no matter how much capital is invested, it will not boost copper production."[22] In order to secure an immediate supply of silver, the domain had to sell 20,000 *koku* of rice, which it had initially planned to ship to Kamigata, within and around the domain. The position was critical: "the emergency situation ha[d] to be taken care of."[23]

Hence, the expulsion of Nagahamaya resulted in the stagnation of silver supply to the copper mine. Directly after expelling Nagahamaya, the Akita domain signed a yearly contract with other copper merchant houses and on 1763/4 long-term contracts were signed with four of these. However, by 1763/11, there had been no improvement in the silver supply situation: "the supply [of silver] to the copper mine is supposed to be on a monthly basis from copper merchant houses in Kamigata but in recent times supply has been irregular and disruptive to the everyday life of the mine workers."[24]

In 1764, according to "*Dōzankata kudashigin hankō*" 銅山方下銀凡考 [the estimate about a copper mine management fund] accounts showed that there was a net shortfall of about 468 *kanme* of silver.[25] Also, the domain had to procure 50,000 *kin* of copper in order to fulfil the agreed quota as indicated by "the supply fee for 500 *ko* of copper shortage to Osaka." In addition, of the 300 *kanme* of silver which was to be used for copper mine management, 150 *kanme* was transferred to Edo and so the silver supply to the copper mine declined

21 上方表御仕送不相違、品々御金繰相違之事のみ出来、上方下銀順達之程不相見得候、左候得は（中略）当春より一向山本へ御渡銀無之候故、此節不被相渡候へば、必定騒立に相及. *Ishii Tadayuki nikki*, 1763/3/23 (Hōreki 13).

22 銅山之儀は一日相後候而は、莫大之出劣に相成、先々程いか程之金銀被入置候ても、出銅之助に不相成. *Ishii Tadayuki nikki*, 1763/3/23 (Hōreki 13).

23 火急之御難事御凌之儀御専務之儀. *Ishii Tadayuki nikki*, 1763/3/23 (Hōreki 13).

24 銅山御仕入銀、上方銅問屋より月割にて指下候筈之所、近来不順達に付、山内日用甚指支候. *Ishii Tadayuki nikki*, 1763/11/6 (Hōreki 13).

25 *Ishii Tadayuki nikki*, 1764/1/19 and 21 (Meiwa 1).

further.[26] A knock-on effect of this was that silver payment to the *tedai gashira* 手代頭, (manager) of the Ani copper mine, was also reduced in that year.[27]

In 1764/5, lead was pawned in order to secure silver supply to the copper mine,[28] but the financial situation at the mine still did not improve. A *tedai* of the copper mine reported on 5/26 that "at Mt. Osawa (Osawa yama 小沢山) within the Ani copper mine, management cannot be sustained unless around 80 *kanme* of silver is transferred."[29] On 7/4 Ishii wrote in his diary that the *tedai* reported as follows:

> From Mt. Osawa within the Ani copper mine, two clerks hastily returned to Akita castle town. Due to the failure of silver supply to the copper mine, *kanako* かなこ (managers of a small mining area) and *kanahori* 金堀 (miners) in every part of the copper mine and labourers are angry. Around 400 labourers who are engaged in cutting down bamboo have gathered with a large number of people, old and young, men and women, and have rushed into the domain's administration house to seek financial support. The mob cannot be brought under control.[30]

Confronted with this crisis at the copper mine, the Akita domain had no option but to reinstate Nagahamaya. On 1764/8/14, a *karō* queried *motokata bugyō* on the subject of how to cope with the crisis and a discussion was held among *motokata bugyō* on 8/21. In the course of this discussion the reinstatement of Nagahamaya was strongly recommended.[31] On 1765/4/16, the reinstatement was planned through the mediation of Tamekawa Jin'emon 為川甚右衛門, although there is no evidence that the plan was put into effect.[32]

---

26 *Ishii Tadayuki nikki*, 1764/4/10 (Meiwa 1).

27 *Ishii Tadayuki nikki*, 1764/2/29 (Meiwa 1).

28 *Ishii Tadayuki nikki*, 1764/5/22 and 26 (Meiwa 1).

29 小沢斗にて凡八拾〆目程の御銀高只今不被相渡候ては崩立. *Ishii Tadayuki nikki*, 1764/5/26 (Meiwa 1).

30 小沢より御手代両人早追にて罷登、御仕送御沙汰無之付枝山共にかなこ金堀大に及迷惑、既に騒立の勢に御座候（中略）竹切の面々四百人程山々御台所へ手廻、老若男女多人数相詰、御助成被下度旨申出候、何共難取鎮候
小沢より御手代両人早追にて罷登、御仕送御沙汰無之付枝山共にかなこ金堀大に及迷惑、既に騒立の勢に御座候（中略）竹切の面々四百人程山々御台所へ手廻、老若男女多人数相詰、御助成被下度旨申出候、何共難取鎮候. *Ishii Tadayuki nikki*, 1764/7/4 (Meiwa 1).

31 古問屋共被相復候儀を御専一ニ被成置候外有之間敷. *Kokuten ruishō*, 399.

32 Until 1763, Tamekawa Jin'emon was a civil servant at the Osaka Copper Trading House (*Osaka dōkaisho* 大坂銅会所), the precursor of the Osaka Copper Agency, as a substitute

2.3 *Reinstatement of Nagahamaya*

The major goal of reinstating Nagahayama and hence restoring the old system was to overcome the crisis in copper mine management. To achieve this it was necessary to improve and stabilize the monthly supply of silver currency to the mine:

> Capital requirements for copper mine management for the entire year are enormous. Thus, the situation at the copper mine will not improve until there is a contract with a copper merchant house which has a stable capital base in Osakaomote 大坂表 and until there are regular monthly transfers of silver money by the middle of each month.[33]

Akita's *motokata bugyō* criticized the fact that there was no sign that *motokata bugyō* Tan Sōjūrō was endeavouring to reinstate Nagahamaya.[34] The attempt at his reinstatement and criticism of the current system had resulted in the enactment on 1764/9/11 of an amendment aimed at financial reform in Akita, Edo, and Kamigata.[35] On the following day, Ōta Ichibei 太田市兵衛 was

---

guardian for a minor, Tamekawa Gorōbei 為川五郎兵衛 of the Dutch Inn in Osaka. In a historical document of the Akita domain, it is recorded that a large debt was owed to Tamekawa Jin'emon as of 1764/8: "the Akita domain's debt concerning copper is not small. Most of the debt is owed to Tamekawa Jin'emon" 銅方借銀不少銀高ニ相成、其内多分為川甚右衛門より之御取替と相見得候. *Kokuten ruishō*, 432. However, it appeared that this *dōgata shakugin* (debt) financed by Tamekawa was the money which he had received from the Shogunate in order to procure copper for overseas export from copper mines other than that in Akita. Tamekawa made an illegal loan to Akita as one can see in the statement "it seems that such a debt was undertaken by Jin'emon and that he secretly loaned it to the Akita domain." 右銀は甚右衛門引請内分秋田方へ貸渡置候趣にて. *Daiisho*, 62. The illegal loan amounted to 472 *kan* 500 *me* and the Shogunate ordered Tamekawa to repay it from 1764 within eight years. In reality the Akita domain paid annual instalments to Tamekawa. It seems that the reinstatement of Nagahamaya was not realized through Tamekawa's financial support because his illegal loan came to light.

33 山本壱ヶ年之御手当銀莫太之御入目ニ候故、於大坂表丈夫之問屋被立置、月割銀は其月中旬迄ニ無相違相達候御年賦ニ無之候而ハ、山本之難渋相止申間敷候. *Ishii Tadayuki nikki*, 1764/8/22 (Meiwa 1).

34 "There is no indication that the previous *kuramoto* has been reinstated." 一向御蔵元共へ被相復候事等不相見. *Ishii Tadayuki nikki*, 1764/8/22 (Meiwa 1).

35 *Ishii Tadayuki nikki*. At this time, since the group consisting of *machi-bugyō* (town magistrate), *jisha-bugyō* (magistrate for temples and shrines), *kanjō-bugyō* 勘定奉行, and *motokata-bugyō* was not considered supportive, Terasaki Yadayū 寺崎弥太夫, Mutō Shichidayū 武藤七太夫, Ishikawa Jōzaemon 石川又左衛門, Iwaya Yahei 岩屋弥兵衛,

appointed as "the person in charge of reform."[36] In the meantime, on 10/4, the *karō* Ishizuka Ichinokami and Imamiya Daigaku 今宮大学 were suspended from their duties.[37] As "their performance was judged to be slipshod"[38] and "Ichinokami had been on a mission since 1761 and his management was extremely bad,"[39] the two were held responsible for the financial deterioration at the copper mine after the expulsion of Nagahamaya in 1761.

The copper mine continued to deteriorate to the extent that "[the situation] is immensely difficult due to inadequate provision of capital to the copper mine [...] [and so bad that] the streets are empty in Ginzanmachi 銀山町 and Mizunashi 水無."[40] A demonstration was organized by labourers when: "around 800 miners who work at the Ani copper mine stayed on the mountains and spent the entire night by their fires. They brought wooden Buddha statues and so on and cut them into pieces for fuel."[41] Without being able to see a gleam of hope for the management of the copper mine and without a guarantee that Nagahamaya could be reinstated, the Akita domain decided to accept the proposal of Itaba Zonshū 伊多波存舟 and promote reform by abandoning, on 1764/12/11, the direct management of the copper mine (*jikiyama*) that had been in place continuously since 1701.[42] However, outwardly the copper

---

Shinda Giemon 信太儀右衛門, and Kataoka Shijūrō 片岡七十郎 were given responsibility. Moreover, on 9/15, the task of *yamamoto goaratame* 山元御改 (mine reform) was assigned to a mine officer. According to the entry from 1764/9/21 in *Satake kafu* financial reform was proposed when "Satake Yoshiatsu 佐竹義敦 entered a *zanoma* (the room where a *daimyō* meets a *kajin*) to make an announcement to *karō*/*kajin*." This marked the initiation of a substantial reform.

36 *Ishii Tadayuki nikki*, 1764/9/21 (Meiwa 1). Moreover, in the context of this reform Akiyama Chōemon 秋山長右衛門 and Yashiro Yadayū 八代弥太夫 were reappointed to *motokata-bugyō*, Onodera Keinosuke 小野寺桂之助 was promoted from *jisha-bugyō* to *karō*, and Kawai Gen'emon 川井源右衛門 was appointed to *motokata-bugyō*.

37 *Ishii Tadayuki nikki*. In addition to the two *karō*, *karō* Hikita Kyūdayū 匹田久太夫 was discharged from his position on 1764/10/13 and *motokata-bugyō* Tan Sōjūrō, who accompanied Ishizuka 石塚 to Kyoto, was placed under house confinement on 1765/12/22 (Meiwa 2) (*Satake kafu*, 1765/12/22).

38 一統勤方忽の様に被思召. *Ishii Tadayuki nikki*, 1764/10/4 (Meiwa 1).

39 市正巳年（宝暦11年）より登坂にて取扱候事甚不宜. *Ishii Tadayuki nikki*, 1764/10/17 (Meiwa 1).

40 山元へ御仕送不参甚及困窮、銀山町・水無往来も無之程の様子. *Ishii Tadayuki nikki*, 1764/9/15 (Meiwa 1).

41 惣山之金堀共八百人余、山々に罷在、今朝迄火を焚終夜居候、卒塔婆仏木などの類を持来打わり焚候. *Ishii Tadayuki nikki*, 1765/2/14 (Meiwa 2).

42 "On the surface, it has the form of *jikiyama* but in reality, based on Itaba Zonshū's proposal, there was an order to make the mine *ukeyama* or a mine under contract

mine remained under the direct control of the Akita domain until the following year, as the domain believed that the transition of the copper mine to *ukeyama* (mine under contract management) would impede the reinstatement of the merchant house.[43] At the same time, with a letter written by the domain's chief and a document drawn up by *karō* Ōtsuka Kurōbei 大塚九郎兵衛 and Okamoto Matatarō 岡本又太郎, *motokata bugyo* Ōta Kuranojō 太田内蔵丞 and Kawai Gen'emon 川井源右衛門 were in the process of negotiating the reinstatement of Nagahamaya in Kamigata.[44]

The Akita domain succeeded in reinstating Nagahamaya soon after the Ani copper mine had been transformed into *ukeyama* on 1765/3/19. According to the diary of *karō* Tsuchiya Tomotora 土屋知虎, the domain assessed Nagahamaya as follows:

> Since 1761 we made every effort to implement the reform [...] and immediately tried to fulfil the quota. However, Takaoka, Yamashita, and others became increasingly powerless with each passing year and could not handle services as *kuramoto*. It was a precarious time for our domain. In these difficult times Nagahamaya was cooperative and strove on, handling both copper and rice shipments, and then annually supplying an enormous amount of capital to fulfil the quota. [...] Recently Nagahamaya has also sold lead and timber produced in Akita. Furthermore, they have dispatched both their *myōdai* 名代 (agent) and *tedai* to the Akita domain,

---

management." 表向は御直山形にて御内々伊多波存舟所存の通受山に可被仰付事. *Ishii Tadayuki nikki*. On 1764/10/17 (Meiwa 1) the first proposal had been put forward by Itaba.

43 "When I asked about the decision to change the management system to *ukeyama*, [I learned that] the decision was taken not to make it a *ukeyama* but to continue to treat it as *jikiyama* for a while since the change will be an obstacle to the reinstatement of the merchant house." 受山之義ニ決着之趣御伺申上候得共、問屋本なりニ被相復候所差障ニ付、無拠暫請山之仕法相止、是迄之通御直山御取扱之儀ニ相決. *Kokuten ruishō*, 439. For details about the *ukeyama* reform of the Akita copper mine, see Sasaki, "Ani dōzan no keiei: Kansei kaikaku o chūshin ni."

44 *Kokuten ruishō*, 75–76. The letter from *hanshu* (*daimyō*) states: "last year in 1761, *karō* in the Akita domain reformed relationships with merchant houses in Kamigataomote. The expulsion of [Nagahamaya]—the merchant house and at the same time *kuramoto*—[...] was extremely disloyal and disappointing." 去ル宝暦十一巳年家老共大坂表仕向令一変、自分共（問屋・蔵元）召放候始末（中略）甚失信義候次第不本意之至ニ候. In this statement *daimyō* first apologizes to Nagahamaya for the erroneous decision by *karō* and others and calls for Nagahamaya to be reinstalled as copper merchant for the domain.

> and the domain has given them extensive control over the transport and handling of rice beyond the bounds of the domain. Nagahamaya was also requested to increase silver transfer to Edo [...] Nagahamaya's efforts are remarkable in this period when not so many *kuramoto* are available.[45]

After his reinstatement, Nagahamaya alone handled the domain's copper, rice, lead, and timber. Based on the account book of 1771, in that year the domain's debt to Nagahamaya amounted to 3,193 *kan* 貫513 *monme* of silver money.

2.4 *Summary*

The Akita domain experienced severe financial difficulties after it lost the financial support of Osaka copper merchants such as Nagahamaya, Yamashita, and Takaoka. Every effort was made to cope with the crisis, including the reinstatement of Nagahamaya and the surrender of direct control of the copper mine. Consequently, financial recovery was achieved after 1766. As a copper merchant and *kuramoto*, Nagahamaya contributed significantly to the recovery by handling all copper and rice.

I would question Sasaki Junnosuke 佐々木潤之介's interpretation that the Shogunate's copper monopoly triggered the monopolization of the copper mine in the Akita domain by transforming the Ani mine from *jikiyama* into *ukeyama*. In the light of the records of the domain's officers such as *Ishii Tadayuki nikki* 石井忠運日記 [Ishii Tadayuki Diary], it seems that the transition was actually part of the Akita domain's effort to recover from the financial crisis that started after the expulsion of Nagahamaya.

3 Copper Merchant House and Income from Copper Sales

As indicated in table 3.1, copper sales of the Akita domain, which had deteriorated after the break with Nagahamaya in 1762, increased following his reinstatement in 1765 and remained stable thereafter. However, this does not mean

45 巳年（宝暦11年）以来御違筋之御取扱ニ異念無之偏ニ御改之思召ニ奉向、（中略）即時より御用向出情相勤候処、高岡・山下義者追年及無力、御蔵元御用相勤兼、誠ニ手薄キ御時節有之候得は、長浜屋深切之所存相尽、御廻銅・御廻米共壱人ニ而引受公私ニ付重キ御用向莫大之出金無御指支年々出情仕候（中略）、別而近年来御領中産物鉛・材木等一手ニ御指向、猶又名代・手代共御当地江指下シ、沖出米をも一統ニ被相任候而、江戸御仕送銀相増御頼被成置、（中略）偏ニ御蔵元手薄キ御時節、長浜や格段之出精無際限次第ニ御座候（後略）. *Tsuchiya Tomotora nikki*, 1771/9/13.

that the financial situation of the copper mine also remained stable. Several years after Nagahamaya had been reinstated the mine was again hit by a financial crisis. Due to the deterioration of Nagahamaya's financial power, a large amount of income from copper sales had to be transferred to Edo instead of Akita and this led to a shortage of silver money for the Akita copper mine. Nagahamaya's financial power had declined because of the stagnation of rice shipments from Akita to Osaka.

In the following I examine the sales of copper for the domain by analyzing the records of gold and silver money movement from Osaka, where the copper was sold, to Edo and Akita. According to "Ginkoku oyoso sashihikisho" 銀穀凡指引書 [the balance sheet for silver and rice] which was submitted to the domain's chief retainer on 1773/5/3, the revenue and expenditure of *nengumai* 年貢米 (annual rice tax) were as follows:[46] of the total revenue of 134,810 *koku* consisting of (1) 99,300 *koku* of *kyūchimai* 給知米 (payroll rice to be supplied to a retainer) (rent from a retainer), (2) 33,320 *koku* of *kuramai* 蔵米 (stored rice), and (3) an additional 13,329 *koku*, a total expenditure of 94,020 *koku* consisting of (4) 33,300 *koku* of *fuchimai* 扶持米 (rice allowance), (5) 45,720 *koku* of rice for shipment to Osaka (including rice as shipment cost), and (6) 15,000 *koku* of *mononarimai* 物成米 (annual rice tax) for the branch family Satake Ikinokami 佐竹壱岐守 was disbursed. The rice for shipment to Osaka (including rice as shipment cost) of (5), which was about 33.5 percent of rice revenue, was transferred from *kyūchimai* of (1). Additionally, 762 *koku* of rice for delivery to Hamagata 浜方 (rice to be returned to Dōjima Hamagata 堂島浜方) was also shipped to Osaka. Thus a total of 46,482 *koku* of rice was shipped to Osaka.

With regard to the Akita domain's silver money income, 1,783 *kanme* of silver was generated in Kamigata from rice and timber sales. Further income derived from *deiriunjōgin* 出入運上銀 (tax levied on ships arriving in port) from ships bound for or coming from present-day Hokkaidō and *okideyakugin* 沖出役銀 (tax levied on unloaded materials at ports within the Akita domain). This amounted to about 3,308 *kanme* of silver in total.[47] The estimated budget of the domain's residence in Edo in 1773 was 2,524 *kanme* of silver. Of this, 1,980

46 *Kokuten ruishō*, 155–60.

47 Since the beginning of the early modern period such silver income had been a special characteristic of the Akita domain, which owned one of the richest silver and copper mines in the nation, and had important ports such as Tsuchizaki 土崎 and Noshiro 能代 at which ships bound for present-day Hokkaidō frequently called. For the Akita domain's finances at the beginning of the pre-modern period, see Yamaguchi, "Akita-han seiritsu-ki no hanzaisei."

kanme of silver was to be transferred from Osaka to Edo and 544 *kanme* from Akita. The amount of silver which was to be transferred from Akita accounted for ca. 22 percent of the total budget in Edo. The domain could procure 1,845 *kanme* of silver in Osaka by selling Akita rice (1,548 *kanme*), by selling timber (35 *kanme*), by selling rice which was initially intended for daily consumption at the mine in Akita (200 *kanme*), and others (62 *kanme*). There was still a shortfall of 679 *kanme* of silver to meet the budget in Edo. Out of this 679 *kanme*, 544 *kanme* was transferred from Akita to Edo. Nevertheless, there was still a deficit of 135 *kanme*.

3.1 *Income from Copper Sales Transferred to Edo*

Nagahayama's key role in the three-sided relationship between Akita, Osaka, and Edo is attested in Ishii's diary. As a *motokata bugyō*, Ishii had to make frequent visits to Akita castle town, the Ani copper mine, Edo, and Kamigata. According to his diary, for a period of three years he resided either in Edo or Kamigata throughout the year; in 1766 he resided in Edo and in 1770 and 1771 in Kamigata. During these years, he recorded the details of silver transfer from Akita and Osaka to Edo.

The 1766 diary seems to be the only one that records the entire annual silver transfer to Edo. In that year, remittance bills[48] to Edo amounted to 43,175 *ryō* of gold, 80 percent of which (34,725 *ryō*) was transferred from Osaka. The remittances from Akita amounted to 212 *kanme* 654 *monme* of silver and *haifukijōgin* 灰吹上銀 (silver obtained through cupellation which was to be paid to the Shogunate as tax). The remittances from Osaka to Edo were classified, for instance, into *gokyō'ōkata* 御饗応方 (banquet expenses), *gotomoyuki no uchi* 御知行之内 (salary for *kerai* [retainers]), *oyakuba goyō no uchi* 御役場御用之内 (cost of office services), and *gotōyōkata* 御当用方 (daily expenses in the Edo *hantei* 藩邸 (residence maintained by a *daimyō*)), and it can be confirmed that Nagahamaya and Hiranoya Mataemon were the issuers of bills as of 1766/4/11. Also, the gold and silver transfer from Akita was conducted during the eleventh and twelfth months; on 12/23, there was a transfer of gold and silver for special purposes such as tax and *gozenkin* 御前金 (money used by a feudal lord). Remittance bills from Osaka were basically issued with rice shipments to Osaka as collateral. Of the rice shipments to Osaka, which totalled 61,000 *koku*, 31,670 *koku* was supplied to Nagahamaya, 7,600 *koku* to Takaoka Kichizaemon, and 5,000 *koku* to Yamashita Hachirōemon.

---

48 With remittance bills a debtor in place A makes settlement with a creditor in place B not by transporting money from place A to B, but by paying with goods such as rice or timber in place B.

In 1770 when the supply of silver to the Akita domain stagnated and led to the financial crisis there, according to Ishii's diary, between 5/26 and 10/4, Nagahamaya received a total of 1,478 *kan* 691 *monme* of silver from the Osaka Copper Agency for 200,000 *kin* realised from the sale of Akita *yodō* 余銅 (left-over copper) and *saodō* 棹銅 (refined copper bars). Of this, 970 *kan* 295 *monme* of silver money was remitted to Akita, which accounted for 65.6 percent of overall income from copper sales. The remainder was possibly transferred to Edo.

In 1771, more of the income from copper sales was transferred to Edo. While Nagahamaya received 2,020 *kan* 貫 200 *me* 目 of silver from copper sales, only about one-quarter (ca. 505 *kan* 50 *me*) was remitted to Akita. From the date on which Nagahamaya received silver for copper sales from the Copper Agency and the date of remittances/transfers of gold/silver from Nagahamaya to Edo, it can be assumed that the following of the many remittances/transfers to Edo derived from copper sales in Osaka.

1. 1700/5/26: Out of 100 *kanme* of advance payment for 200,000 *kin* of *yodō*, 200 *ryō* of gold money, which was part of a debt of 1,200 *ryō*, was remitted to Edo.
2. 6/12: Out of 100 *kanme* of silver debt for *yodō*, 2,300 *ryō* of gold was remitted to Edo on this day.
3. 8/25: Out of 80 *kanme* of income from copper sales, 1,000 *ryō* of gold was remitted to Edo.
4. 1771/1/9: Out of 170 *kanme* of silver income from 425,000 of *mashidō* 増銅 (additionally-sold copper), 1,200 *ryō* of gold was remitted to Edo on this day and on the twelfth day.
5. 2/3: Out of 250 *kanme* of silver debt for *mashidō*, 2,500 *ryō* of gold was remitted to Edo on 2/5.
6. 7/20: Out of 830 *kanme* of silver debt for copper, 3,500 *ryō* of gold was remitted to Edo on 7/21.
7. 8/20: Out of 250 *kanme* of silver debt for *mashidō*, 1,750 *ryō* of gold was remitted to Edo on 8/22.
8. 9/2: Out of 150 *kanme* of the outstanding silver debt for *mashidō*, on 30 *kanme* of silver was remitted to Edo on 9/3.
9. 10/3: Out of 112 *kan* 168 *monme* of the outstanding silver debt for *mashidō*, 1,050 *ryō* of gold was transferred to Edo on 10/4.
10. 10/15: Out of 109 *kan* 564 *monme* of silver income from sales of *tsumikoshidō* 積越銅 (copper which was planned to be sold as quota for the next year), 1000 *ryō* of gold was transferred to Edo on 10/15.

Of the above, 4 and 9 are recorded as transfer of income from copper sales. It can be speculated that the increase in gold remittances to Edo in 1771 was caused by the weakening of Nagahamaya's financial power.

3.2 *Nagahamaya's Financial Crisis*

In 1770 and 1771 Nagahamaya suffered a financial crisis because of the decline in rice shipments from Akita to Osaka. Around 1766, rice shipments to Osaka had been relatively stable. In 1766, they amounted to 61,000 *koku* and in 1767 to 56,000 *koku*.

In 1768, however, there were difficulties with rice shipments to Osaka. It was stated that "last autumn the crops again failed and so rice cannot be shipped at all this year"[49] and that "though the rice, which should have been shipped last year, and about 70,000 *koku* of rice for shipment cost should be shipped this year, due to the exceptionally high harvest loss last year and the resulting shortage of rice, the price has been rising [...] it was reported that rice shipments are not possible."[50] In 1769, reflecting the inflation of the rice price in Osaka (62–63 *monme* of silver money per *koku* for Akita rice), the domain ordered Nagahamaya's *tedai* Takemura Kyūbei 武村久兵衛 not to sell 20,000 *koku* of rice within Akita but to sell it in Osaka. In the same year, Osaka requested Akita to ship an additional 10,000 *koku* of rice over and above the fixed quota of 40,000 *koku*. Since in 1770/3, Nagahamaya submitted an acknowledgement letter confirming that "the rice supply planned for last year had been completed,"[51] it can be assumed that between 40,000 and 50,000 *koku* of rice was indeed sent to Osaka in 1769.

The shortage of rice for shipment became severe in 1770. According to the entry on 1770/7/3 in Ishii's diary, "due to the shortage of rice,"[52] 3 *kan* 120 *me* of silver was given to forty-three people as "this year workers engaged in transportation at the Akita domain's *kuramoto* in Osaka are having a difficult time due to the shortage of rice for shipment to Osaka."[53]

One can speculate that harvest losses and a delay in the collection of rice tax within the Akita domain in 1769 and 1770 led to the stagnation of rice

---

49 去秋も作至而不熟ニ付今年一円廻米難相成. *Kokuten ruishō*, 30.

50 去年中為御登不足米今年江被指延候分とも運賃米凡七万石程被指登候筈之処、去秋作毛不熟幾年ニも無之毛引高多全体米不足にて直段次第引上候付(中略)、御廻米難被成段先達而上方へ申遣候（後略). *Kokuten ruishō*, 30.

51 去年御米調方御皆済. *Ishii Tadayuki nikki*, 1769/2/2, 1769/6/26, 1770/3/1 (Meiwa 6–7).

52 今年為登米不足. *Ishii Tadayuki nikki*, 1770/7/3 (Meiwa 7).

53 今年為登米不足、秋田蔵仲仕難儀. *Ishii Tadayuki nikki*, 1770/7/3 (Meiwa 7).

shipments from the domain to Kamigata, but there is no evidence of this because records from these years are scant in *Satake kafu* 佐竹家譜 (since it does not include the record of *shikyoku* 史局 [historical compilation department of the domain], which served as a source) and a report on the amount of loss within the domain which was submitted annually to the Shogunate does not exist.

No rice shipments from Akita to Osaka took place in 1771.[54] Around this time we find a number of statements describing the decline of Nagahamaya's financial power. For example, the entry for 1770/9/2 in *Tsuchiya Tomotora nikki* 土屋知虎日記 [Tsuchiya Tomotora Diary] states that the domain's forest office (木山方 *Kiyamakata*) proposes to raise funds from merchants other than Nagahamaya since the latter's financial position was weakening:

> Nagahamaya will only be able to make up its deficit this year if it receives sufficient rice from the domain. But the outlook for next year is uncertain. There is no way that Nagahamaya can recover on its own. Even if the domain was to provide it with about 10,000 *koku* of rice, it would not reverse the financial position of Nagahamaya. In order to do so, there is no way other than to send 20,000 or even 25,000 *koku* of rice to Nagahamaya. Nagahamaya's business performance was so dire that the merchant house itself even asked the domain to look to other merchants for fund procurement, otherwise the domain's forest office would also have financing difficulties.[55]

Further details come from Ishii's diary. Between the eleventh and twelfth months of 1771, Nagahamaya attempted to sell 40,000 *koku* of Kobe rice which it had bought before the harvest (spring) of that year. Nagahamaya expected to make a substantial profit by selling this rice at a high price during the season when supply was scarce, but the strategy was unsuccessful. Ishii expressed his deep shock at Nagahamaya's predicament.[56] There is no statement about the

54 "There is no rice shipment [to Kamigata]." 今年一切無回米之. *Ishii Tadayuki nikki*, 1771/9/30 (Meiwa 8).

55 今年不足之所は御渡米被成下相済可申にも可有之候へ共、来年御続方一向不相み得、源左衛門手元も致方無之事に候、来御廻米届壱万石位にて可参事にも無御座候故、壱万も壱万五千も御増米可被指登致候外無之候、右に付木山方等仕入銀に行支候故、別人へ被仰付候様仕度段も申上候事に御座候（後略）. *Tsuchiya Tomotora nikki*, 1770/9/2 (Meiwa 7).

56 （前略）先売米兵庫の方も手切の挨拶にて、所詮長浜屋手元にて出来不申段源左衛門申聞候由驚入申候. *Ishii Tadayuki nikki*, 1771/12/21 (Meiwa 8).

effect on Nagahamaya's finance of the decline in rice shipments from Akita to Osaka. In his letter to Ishii, however, Nagahamaya alluded to the enormity of his misfortune at this time: "As for ourselves, if rice cannot be sold at a higher price, we have no choice but to go bankrupt within the year. Not only are we unable to make various necessary transfers, but we cannot even pay off debt."[57] Furthermore, his financial distress was seriously affecting the domain's finance management as the feudal lord Satake Yoshiatsu 佐竹義敦 warned on 1772/1/13: "in recent years, [since the domain owes large debts to the *kuramoto*], the *kuramoto* in Osaka [Nagahamaya] has a shortage of collateral with which to raise funds. Thus the debt of the *kuramoto* is growing and he has no financial power."

To conclude, the increase in silver money transfer from Osaka to Edo in 1771 was a manifestation of the weakening of Nagahamaya's financial muscle due to the stagnation of the Akita domain's rice shipments to Osaka, and also of the domain's attempt to cover expenses in Edo, specifically, to repay its debt to merchants in Edo. Such a measure to cover expenses in Edo by transferring a large part of the income from copper sales in Osaka inevitably triggered a delay in the monthly transfer of silver money to the Akita copper mine and made the management of the copper business increasingly difficult. The management crisis due to inadequate fund procurement is indicated by the fact that "the mine was already in a state of *yasumiyama* 休山 (mine at rest)."[58] The following is also stated in the 1774/5/22 entry of *Okamoto Matatarō Mototaka* 岡本又太郎元貴 *Diary*:

> I have been informed several times that rice shipments from Akita to Kamigata are stagnating and thus the silver money supply to the domain's mine has not taken place. Furthermore, our copper merchant and *kuramoto*, Nagahamaya, even has difficulty in operating normally and obtaining working capital. I am deeply concerned by the shortage of silver money supply at the mine in recent years. Lately the situation has become even worse and there are several points that are unacceptable.[59]

---

57 （前略）私方も右の米売不申時は年内切潰申外無之、外々御仕送は勿論、借銀の歩払小払出来不仕候. *Ishii Tadayuki nikki*, 1771/12/21 (Meiwa 8).

58 既に休山程ニ相成候故. *Kokuten ruishō*, 491.

59 大坂より山方下シ銀御回米為御登相違ニ付相滞候よしは上方より具々申来候得共、蔵本銅問屋両様一手ニ長浜屋取担候ニ付、御当用之引当難渋ニ付、山方之下シ銀相違之事ニ成候は先年より之深キ障りニ候、当時はヶ様之形猶以大ニ差障り而已多く難心得事共種々有之候. *Okamoto Matatarō Mototaka nikki*, 1774/5/22.

The Shogunate's provision of financial support in the amount of 10,000 *ryō* of gold for copper shipments to Osaka in 1773 can be attributed primarily to the financing need which resulted from the deterioration of Nagahamaya's financial strength.

3.3 *Summary*

Although the Akita domain succeeded in reinstating Nagahamaya, its rice shipments to Osaka stagnated and so the financial power of Nagahamaya weakened. As a result, the income from copper sales in Osaka was not invested in the copper business in Akita. Instead, ca. three-quarters of it was transferred to Edo in 1771. This demonstrates that although the copper price was fixed under the Shogunate's copper monopoly, the income from copper sales was used to secure the financing of the domain's residence in Edo, which was normally covered by the sales of Akita rice in Osaka. That copper fulfilled this function is clearly evidenced in 1771 when Akita rice was not sold in the Osaka market.

4 Conclusion

The Shogunate fixed the price of the copper to be exported abroad at a low rate since it was necessary for the Shogunate to offer Japanese copper to overseas merchants at a low price. The statement "copper sold in Nagasaki [for overseas export] is just like a tax on mines, thus copper should [all] be transported to Osaka [for that purpose]"[60] embodies the logic of the Shogunate. Thus, the interpretation is generally accepted that the mines could not receive payment for their copper that was commensurate with production costs and as a result copper production stagnated. In the case of the Akita domain, however, the Osaka copper merchant house Nagahamaya played the significant role of compensating for this shortfall. In other words, it was the financial relationship between Nagahamaya and the domain that determined its copper production and shipments to Osaka.

By re-evaluating the significance of the Shogunate's copper policy with particular reference to the role of the Osaka merchant houses,[61] it becomes

60 長崎廻り銅ハ諸国山々公役同前ニ而大坂え差廻し. *Kiyō gundan*, 312.

61 Of the studies on the Shogunate's policy on copper production and the management of the copper mines that followed Ishii Takashi's, that of Sasaki Jun'nosuke illuminated the problems inherent in the Shogunate's policy most clearly. However, concerning the issue of the Akita domain's finance and its copper mine management during the Hōreki and Meiwa period (1751–1769), he argued that the domain's financial policy had nothing to do

apparent that copper functioned as more than a *kubutsu* 公物 (public property or tax) for the domain. Specifically, the domain successfully obtained a loan from Nagahamaya using 830 *kanme* of silver money as collateral, which was to be paid by the Shogunate for the copper produced in the coming year.[62] In this way, once the Shogunate had established its monopoly over copper, copper became a commodity which could generate a large amount of income. Payment for the copper, which was made to the domain by the Nagasaki clearing house for foreign trade accounts (*Nagasaki kaisho* 長崎会所) via the Osaka Copper Agency, became collateral for the Osaka copper merchant house. The Akita domain attempted to gain access to the capital of the Osaka merchant house by making full use of the copper's potential as collateral. Furthermore, the domain occasionally used the income from copper sales to fill financial gaps. By examining the relationship between the Shogunate, the Akita domain, and the Osaka copper merchant house in the 1760s and 1770s, we were able to find evidence of the above-mentioned function of copper which has hitherto not been discussed. Copper in early modern Japan, which has so far been seen as public property or tax as indicated in the statement above, should now be re-evaluated as a commodity.

---

with the capital of Osaka merchant houses and interpreted the *ukeyama* reform (transformation of the Ani copper mine into a mine under contract management) in the Meiwa period as the result of the domain's financial crisis and dependence on domestic capital. Consequently, he did not probe the relationship between Osaka copper merchant houses and the domain's financial policy, or the influence of the copper merchant houses on the domain's copper production and shipments. See Sasaki, "Kinsei sandō seisaku ni kansuru ichikōsatsu."

62 長浜屋手代十右衛門・彦兵衛・孫兵衛旅宿へ扣候て、五郎右衛門立会、江戸表御差支并御下国御用三千両指向候義、最早時節指掛候間、此節より段々仕送候様致度、右調達之儀ハ、銅御前渡八百三拾〆め引当候て也. *Ishii Tadayuki nikki*, 1771/3/18 (Meiwa 8).

CHAPTER 4

"There's Method in the Madness"*: A New Approach to the Early Modern Sino-Japanese Copper Trade

Keiko Nagase-Reimer 永瀬ライマー桂子

1 Introduction

In the middle of the Tokugawa period (1603–1868) copper was Japan's main export good. Its pricing, however, looks 'irrational' at the first glance: The Shogunate institution called Copper Agency (*dōza* 銅座) bought copper for export from the mines at a very low price. This price was fixed and in the late Tokugawa period it could not even cover the production cost.[1] Furthermore, the *Nagasaki kaisho* 長崎会所 (Nagasaki clearing house for foreign trade accounts), also a Shogunate institution, exported copper at fixed low prices, which were even lower than the price the Copper Agency paid to the mines.

Until today many historians try to give an explanation for this copper pricing. They agree that this seemingly irrational copper pricing was financially profitable for the Shogunate (*bakufu* 幕府): The *Nagasaki kaisho* sold copper at a low price to foreigners. In return, the *kaisho* could obtain foreign goods also at a low price. With the margin for selling these foreign goods, the *Nagasaki kaisho* could obtain enough income and with this income, it could compensate the loss made by selling copper.[2]

The opinions of historians are, however, split regarding who bore the financial burden. Some historians argue that the mines suffered most from this seemingly irrational copper pricing. According to them, the mine managers accepted the low copper price because their main purpose was not to make a profit by copper production, but to serve the Shogunate by producing copper and paying business tax (*unjōkin* 運上金) to the Shogunate.[3] Other historians believe that the copper consumers within Japan bore the financial burden. According to them, the copper price for domestic consumption was set so high that the mines could compensate the loss incurred by selling copper for export

* Shakespeare, "Hamlet" act 2, scene 2, 43.

1 Sumitomo iinkai, *Sumitomo Besshi kōzanshi*, 242.

2 Nakamura, *Kinsei Nagasaki bōeki-shi no kenkyū*, 375.

3 Nagazumi, "Osaka dōza," 418–49; Yamaguchi, "Kingin-zan no gijutsu to shakai," 160–61.

 | DOI 10.1163/9789004304512_005

purposes with the profit made by selling copper on the domestic market. Therefore the mines made no financial loss. Instead, the domestic consumers were forced to buy copper at a high price.[4]

The above explanations are all based on the view that the Shogunate strictly controlled the copper production and price setting with its absolute power. All the other parties involved were forced to obey the Shogunate's decision. This paper aims to give another explanation to this seemingly irrational copper pricing which in turn provides a different interpretation of the role of the Shogunate.

2 Historical Background

In the course of Japanese history there were three boom phases in mining. The first boom took place in the 7th and 8th centuries in a period of state-building and intense contact with the continent. In this period mining was developed to provide metals for the construction of palaces and temples.[5] The second boom occurred from the mid-15th to the mid-16th century, starting at the end of the so-called Warring States Period (*sengoku jidai* 戦国時代 1467–1573). The third boom ensued in the wake of technological modernisation in the late 19th century.[6]

The second mining boom was triggered by an expanding circulation of gold and silver, the metals regarded as the most valuable in the society of the Warring States Period. Gold and silver were used as rewards and payments for soldiers and hence began to circulate as monies. For the maintenance of their armies, the feudal lords, the *sengoku daimyō* 戦国大名, had a high demand for precious metals, and therefore were eager to develop gold and silver mines within their territories.[7] Silver was especially abundant and began to be exported. Kobata Atsushi 小葉田淳 (1905–2001), the scholar who has most

4 Sasaki, "Dōzan no keiei to gijutsu," 184–85; Shimada, *The intra-Asian trade in Japanese copper by the Dutch East Company during the eighteenth century*, 57–59.

5 Mining began relatively late in Japan. This was due in part to the fact that political and trade relations permitted the importing of refined metals from Korea until the middle of the 7th century. The author thanks Nanny Kim for this comment. See also Murakami, *Kin gin dō no Nihonshi*, 52–54.

6 Kobata, *Nihon kōzanshi no kenkyū*, 3; Murakami, *Kin gin dō no Nihonshi*, 52 and 129.

7 Kobata, *Nihon kōzanshi no kenkyū*, 3.

widely researched the history of Japanese mining, estimates that by the early 17th century exports of silver from Japan amounted to 200 tons annually.[8]

However, when gold and silver mining declined in the second half of the 17th century, copper production increased dramatically.[9] In 1685 there were altogether fifty copper mines in Japan, and the amount of crude copper collected in Osaka reached ninety million *kin* (54,000 tons) in that year.[10] Copper then replaced gold and silver as the main export good. Copper accounted for about 2/3 of the export amount in currency in 1684.[11]

Copper became such a popular export good because there was a high demand for Japanese copper overseas. Until the Yunnan 雲南 copper mine in the south-west of China was developed in the mid-18th century, China was dependent on Japanese copper for the minting of coins. The VOC (Dutch East India Company) also started to buy copper instead of gold from Japan in the late 1640s and sold it in and around India. There, military goods, coins, household goods, and Buddhist altar equipment were produced using Japanese copper. When the Second Northern War (1655–1660) broke out in Sweden, one of the main copper producers in Europe at that time, the copper market price in Amsterdam rose dramatically. Therefore, the VOC switched from importing sugar from Southern Asia, the market price of which was just beginning to fall, to importing large amounts of Japanese copper. Copper was a perfect substitute for sugar since copper could also be used as ballast on the ships. When the copper price began to fall again in Europe in the 1680s, the VOC continued to buy Japanese copper and sold it once more in India and its surroundings.[12]

Japan's prosperity from the copper export did not last long, though. Although the demand for copper remained high, Japan's copper mines started declining and their production could not keep up with the demand. As a consequence, less and less copper was available for export.[13] The Shogunate tried to recover the level of production by issuing laws to promote the development of new

---

8 Kobata, *Kōzan no rekishi*, 59.

9 Ibid., 117.

10 Ibid., 179.

11 Ibid., 182–83.

12 Wakamatsu, "Kinsei chūkōki no Nagasaki dōbōeki to kokunai sangyō," 292–93; Murakami, *Kin gin dō no Nihonshi*, 142–44; Shimada, *The intra-Asian trade in Japanese copper by the Dutch East Company during the eighteenth century*, 12–28.

13 Wakamatsu, "Kinsei chūkōki no Nagasaki dōbōeki to kokunai sangyō," 292–93.

mines in 1695[14] and 1698.[15] However, this was not as effective as expected. Then the Shogunate changed its policy and started to control the production and circulation of copper.[16] In 1701, the Shogunate established the First Copper Agency (*daiichiji dōza* 第一次銅座) in Osaka 大坂. Crude copper (*aradō* 荒銅) produced all over Japan was collected in Osaka, and its amount was reported to the Copper Agency. At that time Osaka was the centre of commerce in Japan, and the most influential stakeholders for copper procurement such as the association of copper refineries (*dōfukiya nakama* 銅吹屋仲間) and the copper merchant houses (*dōton'ya* 銅問屋) were settled there.

After the establishment of the First Copper Agency there were some changes in the Shogunate policy on controlling copper. The Copper Agency was twice abolished and twice newly organized. Nevertheless, the Shogunate continued to control the production and circulation of copper until the end of the Tokugawa period. Despite the Shogunate's control, the copper production continued to decline and then stagnated from the middle of the 18th century. It is difficult to know the exact amount of copper produced in Japan throughout the early modern period. According to economic historian Shimada Ryūto, it lay at 3,840 tons per year in the 1710's. In 1770's Japan produced about 2,700 tons of copper. In other words, the amount of copper produced was reduced by half within seventy years.[17]

3 The Seeming Madness: The Copper Pricing

Usually the prices of goods increase steadily on the way from production to end consumer. This was, however, not the case with the Japanese copper trade. In the late Tokugawa period the price of copper exported to Chinese merchants was even lower than the production costs. For example in 1791, as shown in figure 4.1, the production cost of 100 *kin* (60 kg) of copper in the Akita 秋田

14 *Ofuregaki kampō shūsei*, 1759, "Kuniguni tokorodokoro kingindōzan korearuwa enryonaku saikutsu no koto" 国々所々金銀銅山有之ハ無遠慮採掘之事 [Order not to hesitate to mine if there are gold, silver, and copper mines anywhere the country].

15 *Ofuregaki kampō shūsei*, 1765, "Goryō shiryō no uchi kingindōzan mitate horimi mōsubeki no koto 御料私領之内金銀銅山見立堀見可申之事 [Order to investigate gold, silver, and copper mines in official as well as private areas].

16 Yoon, *Domain and Bakufu in Tokugawa Japan*, 140.

17 Shimada, *The intra-Asian trade in Japanese copper by the Dutch East Company during the eighteenth century*, 51.

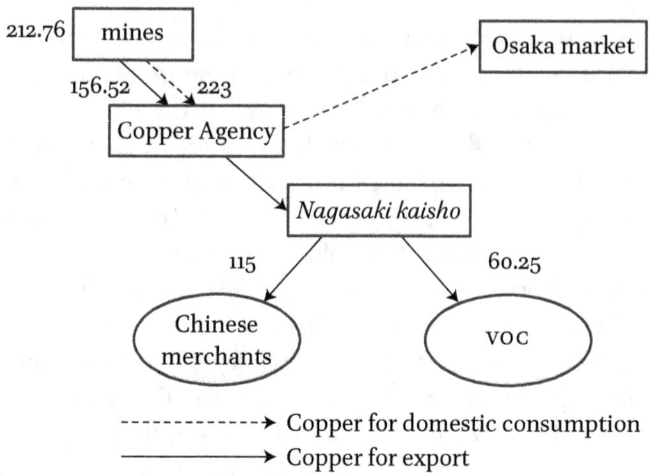

FIGURE 4.1 *Copper prices in 1791 (unit:* monme/*100* kin).[18]

domain's mines was 212.76 *monme* 匁 in silver currency. The Copper Agency bought this copper at two different prices depending on purpose: For the domestic sale, the Copper Agency bought the copper at 223 *monme*.[19] That margin was not very large, but still the mines could make some profit through this business. However, for the copper for export the Copper Agency paid only 156.52 *monme*.[20] In this case, the copper mines obviously made a loss. After refinement in Osaka, the copper for export was shipped to Nagasaki 長崎. There, the *Nagasaki kaisho* sold it to Chinese merchants at an again massively lower price—only 115 *monme*/ 100 *kin*[21]—and for even less to the VOC—almost 50 percent less, about 60.25 *monme*[22].[23] Thus, the *Nagasaki kaisho* made a loss by exporting copper.

18 The copper which the Copper Agency purchased from the mines was crude copper. On the other hand, the copper which the Copper Agency sold to the *Nagasaki kaisho* was refined copper. Therefore we cannot directly compare these prices. However, we can see the tendency. The copper price at which the Copper Agency sold it to the *Nagasaki kaisho* is not available.

19 Akitaken, *Akitaken-shi* 3, 427.

20 Akitaken, *Akitaken-shi* 3, 424; Kudō, *Ani-ai kyōdoshi*, 35.

21 *Daiisho*, 10; Yamawaki, *Kinsei Nitchū bōeki-shi no kenkyū*, 74.

22 Suzuki, *Kinsei Nichiran bōekishi no kenkyū*, 206. "Oranda kata shōbai nimotsu motoharai tō daii yakusho" 阿蘭陀方商売荷物元払等大意訳書 [Documents on costs of goods traded with the Dutch] show, however, that it was 60.25 *monme*. See *Nagasaki kaisho gosatsumono*, 132.

23 Besides to China and VOC from Nagasaki, Japanese copper was exported to Korea through the Tsushima domain during the Tokugawa period. The Sino-Japanese copper trade was

It was the decision of the Shogunate to set the price for the export copper low. In the late 17th century, when Japan still produced copper in abundance, it made a solid profit by selling copper to Chinese merchants at a low price. At the turn of the century, however, the copper price rose on the Japanese market. The Japanese merchants[24] tried to sell copper at a higher price to the Chinese merchants and to the VOC, reflecting the market situation in Japan. However, the negotiation with the Chinese merchants and the VOC was not easy. In 1701 the Shogunate gave up their attempts to raise the price of export copper and instead introduced a system to procure copper for export at a lower price from the mines. The Shogunate declared the copper for export as *goyōdō* 御用銅 (literal translation: copper for official use). The goods for official use usually had the highest priority and their price could be set low. This enabled the Shogunate to procure a certain amount of copper at a low price and to use it according to the priority of the time—for export, for coin minting or to repair the Shogunate's constructions.[25] Even though the Shogunate was able to procure copper for official purposes at a low price, the selling prices to the Chinese merchants and the VOC in Nagasaki were still lower and the fact remains that the state made a loss by exporting copper.[26]

From a modern point of view it is also curious that the selling prices to the Chinese merchants and to the VOC were fixed independently from market price, production costs, or demand: We understand today that the price of a product rises when there is less supply and increasing demand. However, supply and demand had nothing to do with the price for the export copper. It was fixed at 115 *monme*/100 *kin*[27] for over 100 years.

Among the contemporaries, there were also some who wondered why *the Nagasaki kaisho* offered Japanese copper at a too low price. For example in 1733, Nagasaki magistrate (*Nagasaki bugyō* 長崎奉行) Ōmori Yamashiro no kami Tokinaga 大森山城守時長 (1690–1761) and his colleagues suggested to stop selling copper to the Chinese at less than half the Japanese market price.[28] They further suggested that the copper price for official purposes and the copper price for domestic sales (*chiuridō* 地売銅) should be the same.

---

the largest and is therefore the focus of this paper. Regarding the Korean-Japanese copper trade, see Tashiro "Tsushima-han no Chōsen yushutsu dō chōtatsu ni tsuite," 141–208.

24 Before the establishment of the First Copper Agency the copper trade was carried out by copper merchants. See Sumitomo shūshishitsu, *Dai ichi ji dōza to Sumitomo*, 38–39.

25 Imai, "Kinsei Nihon no dō," 24; Sumitomo shūshishitsu, *Dai ichi ji dōza to Sumitomo*, 71; Nagazumi, "Osaka dōza," 410.

26 Yamawaki, *Nagasaki no Tōjin bōeki*, 172–73.

27 *Daiisho*, 10; Yamawaki, *Kinsei Nitchū bōeki-shi no kenkyū*, 74.

28 Iwasaki, "Genbun dōza-ki no Nagasaki kaidō to Nagasaki bōeki," 4–5.

These proposals, however, were not realized because one of the central figures of this proposal died in that year, and Ōmori was dismissed in 1734.[29] Another example is Kudō Heisuke 工藤兵助 (1734–1801), a physician from the Sendai domain and an economist. He referred to the pricing structure in his writing *Hōkoku igen* 報国以言 from the 1780's.[30] There, he inferred that the *Nagasaki kaisho* had kept the low selling price to the Chinese merchants from the times that copper had been abundant because the *kaisho* received bribes from the Chinese and Dutch.[31]

This brings us to the title of this paper, as the seemingly irrational copper pricing of the Shogunate shows some parallels to the protagonist's behaviour in Shakespeare's *Hamlet*. There, the Lord Stewart Polonius is convinced that Prince Hamlet is truly mad. Nonetheless he recognizes some "method" in Hamlet's speech—that is, a kind of reason and order.[32] The copper pricing in early modern Japan certainly looks 'mad' at first glance. However, it, too, had some 'method'. The method seen by historians so far is that the *Nagasaki kaisho* could actually still make a profit with this pricing. The trick was that the loss made by copper export was compensated by the income from selling imported goods to the Japanese merchants.[33] In this way, the Shogunate profited from the seemingly irrational copper pricing. The mines and copper consumers in Japan were simply forced to accept the Shogunate's unreasonable pricing. Thus, they bore the financial burden.

Yet, this paper suggests a different 'method' regarding the irrational copper pricing by analysing the flow of copper, cash, and the imported goods as a whole, as shown in figure 4.2.[34]

In the following sections, the analysis of the price setting will be carried out along the schema depicted in figure 4.2. Section 4 will look at the right-hand side of the schema, starting from the bottom to explain the transactions between the *Nagasaki kaisho* and the Chinese merchants. Furthermore, the distribution of the imported goods within Japan will be examined with the concrete example of the goods brought by a Chinese junk which arrived in

29 Ibid., 4–5.
30 Yamawaki, *Kinsei Nitchū bōeki-shi no kenkyū*, 73.
31 *Tsūkō ichiran* 8, 494–95.
32 Shakespeare, William, *Hamlet* act 2, scene 2, 43.
33 E.g. Nakamura, *Kinsei Nagasaki bōeki-shi no kenkyū*, 375.
34 The intensive discussions about "commodity chain" in the workshop "Sino-Japanese trade in the 18th and 19th centuries: Where was the profit made?" with George Bryan Souza at the Ruhr-University Bochum on January 21st and 22nd, 2010 gave me some ideas regarding this methodology.

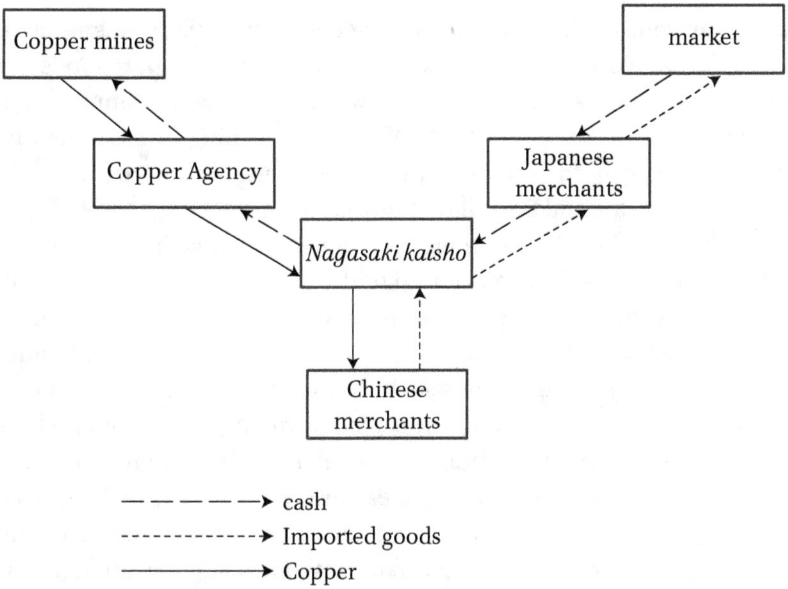

FIGURE 4.2 *Flow of export copper, imported goods, and money within Japan.*

the port of Nagasaki in 1812. Section 5 discusses the various explanations historians have given so far for the pricing of export copper. In section 6, we will look at the left-hand side of the schema and examine the Copper Agency, the copper merchants and consumers. Section 7 examines the copper production. Based on this analysis, I will then suggest a different "method" behind the copper pricing in the conclusion in section 8.

4 Nagasaki Transactions and the Distribution of Foreign Goods within Japan

In the introduction I mentioned that the *Nagasaki kaisho* retained their 'irrational' price structure because the loss was compensated by the income from selling imported goods to Japanese merchants. This chapter explains how this functioned by using concrete examples.

Copper and other export products were gathered in Nagasaki and then exchanged/traded for commodities brought by Chinese merchants or the VOC. The trade in Nagasaki was in principle a compensation trade (*kyūshō bōeki* 求償貿易), a kind of barter trade.[35] In this trading system, the value of import

35 Yamawaki, *Kinsei Nitchū bōeki-shi no kenkyū*, 75.

goods was equivalent to the value of export goods. Thus, there was no cash movement. In other words, the *Nagasaki kaisho* paid for the import goods not with cash, but with Japanese products as shown in figure 4.2. The annual total value of regular trade (*osadamedaka* 御定高, 'trade ceiling') was limited by the Shogunate.[36] In 1791, for example, the trade ceiling was 2,740 *kan* 貫 in silver currency (today around 80 million Euro, i.e. 8 million Euro/vessel).[37] The *Nagasaki kaisho* and the Chinese merchants agreed a 'price' for each good. The Chinese merchants 'sold' goods for up to the equivalent of this trade ceiling. The *Nagasaki kaisho* first subtracted the costs for the stay of the Chinese merchants in Nagasaki from this value. Then the *kaisho* 'paid' the remainder of the total value with Japanese goods such as copper.[38] As mentioned above, the *kaisho* offered copper at a very low price. In return, the *kaisho* obtained the foreign goods from the Chinese merchants also at a low price. Here, the 'price' was nothing to do with the price on Japanese or Chinese markets. They used these 'prices' only to calculate how many goods they could exchange within the trade ceiling. Since this was a barter trade without cash movement, it was a fair trade as long as the parties could exchange goods of equal value in total. Even though the prices of the Japanese goods were too low, the *kaisho* would make no loss as the prices of the Chinese goods were also low. Since there was a trade ceiling, when the *kaisho* and the Chinese merchants agreed to set the prices low, they could exchange a large amount of goods; when they decided to set the prices high, they could only exchange a small amount of goods. The trade became profitable for the *Nagasaki kaisho* when the income from selling the imported goods to Japanese merchants exceeded the expenditure, namely the loss from selling Japanese goods and the operating expenses.

With this in mind, let us examine two concrete examples—sappan wood (*soboku* 蘇木) and cardamom (*shukusha* 宿砂) for medicine—which were carried to Nagasaki by the Chinese junk Yangmao 永茂 in 1812.[39] As shown in table 4.1, the *Nagasaki kaisho* purchased the sappan wood from the Chinese at 0.9 *monme* per item (cost price). The *kaisho* then sold it to Japanese merchants

---

36 Ibid., 74–75.

37 Yamawaki, *Nagasaki no Tōjin bōeki*, 165. This estimation is based on the assumption that one silver *monme* equals 4,000 yen (30 Euro).

38 This applies to the so-called *hon-uri* 本売, the regular trading. However, there were two more trading possibilities: one is the so-called *betsudan-uri* 別段売, the special trade. Here the Chinese were allowed to sell their personal belongings. The other is smuggling, which was of course illegal, but supposedly happened quite often. See Nakamura, *Kinsei Nagasaki bōeki-shi no kenkyū*, 423–24, 484.

39 Nakamura, *Kinsei Nagasaki bōeki-shi no kenkyū*, 490–91.

TABLE 4.1 *Price per item* (*unit*: monme *in silver currency*)[40]

	sappan wood	cardamom
Cost price (*motone* 元値) of the *kaisho*	0.9	4
Sale price (*rakusatsune* 落札値) of the *kaisho*	4.8	7.936
Price in Osaka market (max.)	9.5	10.7
Price in Osaka market (min.)	6.7	7.7
Profit of *Nagasaki kaisho*	430 %	98 %
Profit of Japanese merchants	Between 40% and 98%	Between −3% and 35%

at 4.8 *monme* (sale price). Through this, the *kaisho* made a massive profit of 430 percent. The price on the Osaka market was fluctuating: in 1813, the lowest price was 6.7 *monme* and the highest price was 9.5 *monme*, which means the Japanese merchants made a profit of between 40 percent and 98 percent. It was much less than the profit of the *kaisho*, yet it was quite profitable. However, sometimes the Japanese merchants incurred a loss, e.g. with the cardamom, as the market price in Osaka was quite low, as shown in table 4.1. Yet the *Nagasaki kaisho* still made a profit. In this way, the *Nagasaki kaisho* generated profit from selling foreign goods. This system, which made the *Nagasaki kaisho* profit in any case, was established in the middle of 18th century.[41]

5 'Method' of the Copper Trade

In the last decades, historians have been faced with the challenge of giving an explanation for the low prices at which Japanese copper was sold to foreign merchants. Japanese scholar Ishii Takashi 石井孝 tried to explain the low copper export price with the tribute trade (*chōkō* 朝貢). The tribute trade is a pre-modern trade form in the surroundings of China. The ruler of a neighbouring country (here: the Shogun of Japan) brought tribute (= copper) to the Emperor of China (in this case: the Emperor of the Qing dynasty) to show his respect. In return, the Chinese Emperor would give generous gifts (*kaishi*

40 Nakamura, *Kinsei Nagasaki bōeki-shi no kenkyū*, 490–91.

41 Imai, "Kinsei Nihon no dō," 35.

回賜).[42] Historian and former head of the Sumitomo archive Imai Noriko 今井典子, however, refuted Ishii's explanation based on two arguments: firstly, in the 17th century Japan did make a profit with their export trade. It was only in the second half of the 18th century that the copper trade became a deficit business. Secondly, the Sino-Japanese trade at that time was a private trade, there were no negotiations between Japan and the Qing government and no tribute ceremony.[43]

Ishii also offered another thesis: He explained that the *Nagasaki kaisho* set the copper price so low in order to buy goods from the Chinese at equally low prices and so to make a large profit.[44] This time, historian Yamawaki Teijirō 山脇悌二郎 refuted Ishii's explanation. According to Yamawaki, a large part of the profit of the *Nagasaki kaisho* went to the people in Nagasaki while the Shogunate only obtained the business tax, which was just a small part of their yearly income. He argued that it is therefore not plausible to think that the Shogunate continued to sell copper below its production cost only for this small income and for the benefit of Nagasaki.[45] Yamawaki's argumentation is plausible: at the beginning of the 18th century, the Shogunate received about 70,000 *ryō* in gold currency as business tax from Nagasaki and this sum played a certain role in the financial reform of the Shogunate.[46] In the 1760s and 70s, however, the business tax from Nagasaki only amounted to 15,000 *ryō* and we can indeed assume that this was less significant for the Shogunate.[47]

Yamawaki argued that for the Shogunate, the main purpose of the Nagasaki trade was not the profit, but the exchange of large amounts of goods. Therefore Yamawaki concluded that the Shogunate fixed the copper selling price to foreign merchants low in order to exchange as many goods as possible within the trade ceiling.[48] This explanation is plausible. If profit had been the main purpose of the Nagasaki trade, the Shogunate would have pursued a trade surplus and would not have implemented the compensation trade.

The Shogunate could theoretically also have exchanged larger amounts of goods with Chinese merchants by loosening or even abolishing the trade ceiling instead of fixing the copper price at low. However, in my opinion this option was not taken because the Shogunate regarded the trade ceiling as

42 Ishii, *Bakumatsu kaikōki keizeishi kenkyū*, 235.

43 Imai, "Kinsei Nihon no dō," 34–35.

44 Ishii, "Bakkumatsu ni okeru bakufu no dōyushutsu kinshi seisaku," 3.

45 Yamawaki, *Kinsei Nitchū bōeki-shi no kenkyū*, 73–76.

46 It is based on the data of 1703 and 1704. See Ōno, *Edo bakufu zaisei-shi ron*, 248–50.

47 Ōno, *Edo bakufu zaisei-shi ron*, 311–12.

48 Yamawaki, *Kinsei Nitchū bōeki-shi no kenkyū*, 73–76.

necessary to prevent the outflow of precious metals to overseas: since the second mining boom, Japan had paid import goods with gold and silver. Yet in 1685, the Shogunate had fixed an annual value limit for payments in gold or silver. Above this limit, the Shogunate allowed payments in copper, *tawaramono* 俵物 (direct translation "bale goods", concretely dried sea cucumber, dried abalone, and dried shark's fin), or *shoshiki* 諸色 (kelp, a dried cuttlefish, *tosaka* seaweed, *tengusa* seaweed, etc.). In 1715 Arai Hakuseki 新井白石 (1657–1725), who supported the 6th Tokugawa Shogun Ienobu 家宣 (r. 1709–1712) and the 7th Shogun Ietsugu 家継 (r. 1713–1716), introduced the *Shōtoku shinrei* 正徳新例 (literal translation: New Regulation of the Shōtoku era). This prescribed that import goods were to be paid not with gold or silver, but only with copper, *tawaramono, shoshiki* and others (*shiromonogae* 代物替). Hakuseki compared the mining resources of a country to a person's bones, since both do not grow again once they are lost.[49] Hakuseki was convinced that it was nonsense to exchange such precious resources with goods such as raw silk or sugar which could grow again like human blood, flesh, hair and skin.[50] Obviously, the *Shōtoku shinrei* still allowed the export of unrenewable copper; nevertheless, they limited the amount of copper to be exported.

Hakuseki was not the only one who wanted to limit the export of precious metal and refrain from importing goods such as raw silk and sugar. Hakuseki's idea can also be found in writings by various intellectuals from the Edo period such as *Gendōroku* 減銅録[51] by Hanai Ikkō 花井一好, who was a *yoriki* 与力 (administrative assistant at the Shogunate's office), *Sōbōkigen* 草芽危言[52] by Nakai Chikuzan 中井竹山 (1730–1804), *Uge no hitogoto* 宇下人言[53] by Matsudaira Sadanobu 松平定信 (1759–1829), and *Keisei hisaku* 経世秘策[54] by Honda Toshiaki 本多利明 (1743–1821). We can see their influence in the

---

49 "Hakuseki kengi roku," 242.

50 Ibid., 242.

51 Hanai, "Gendōroku," 376.

52 See vol. 4, "Gaihakugoshi no koto" 外舶互市ノ事 [About trade] in Nakai, *Sōbōkigen*, 362.

53 Matsudaira, *Uge no hitogoto*, 102.

54 Honda, *Keisei hisaku*: 1, 52 and 56. Honda's interpretation is very unique in that the Shogunate prevented the outflow of copper to overseas by purchasing official copper at a low price from the mines. The Shogunate intentionally fixed the official copper price low so that the copper mines would make a loss by their production. According to Honda, the Shogunate aimed to discourage the mines with this, then the mines would stop producing copper, the copper production would decrease and finally the outflow of copper to overseas would be avoided.

fact that in the late Tokugawa period *tawaramono* replaced copper as the main export good and domestic production of raw silk, silk textiles and sugar began.[55]

Another aspect that should not be forgotten when considering the copper pricing is that it was not only the Shogunate that benefitted from it. Apart from them and their institutions such as the *Nagasaki kaisho* and the Copper Agency, the Japanese merchants also favoured the fixed low copper price, as this also meant low cost prices for import goods. The lower the cost price was for them, the higher the margin they could obtain from selling the goods on the markets and consequently the higher their profit. Therefore it must have been desirable for them, too, to keep a trading system in which Japan could obtain foreign goods at low prices.

The *Nagasaki kaisho* sold foreign goods to the Japanese merchants at a high price in order to generate enough profit. The Japanese merchants then tried and often succeeded to sell these foreign goods at an even higher price to the Japanese consumers. However, the Japanese consumers were not forced by the Shogunate to buy foreign goods at exorbitantly high prices. The market mechanisms worked well on the early modern Japanese markets. If the goods were too expensive, the consumers did not buy them and consequently, the Japanese merchants had to reduce the price. Thus, the merchants sometimes incurred a deficit as could be seen with the example of cardamom in table 4.1.

It is certainly true that the Shogunate tried to manipulate the market in order to keep the prices of foreign goods high. The Shogunate limited, for example, the amount of cheaper Japanese goods on the Osaka market in order to sell more expensive foreign goods. However, their attempts were not always successful.[56] Foreign goods for which the Shogunate set the price too high lost their competitiveness on the market.[57] Even for the Shogunate it was not possible to control the prices on the markets and keep the prices of foreign goods high.

55 See Matsuura "The trade in dried marine products from Nagasaki to China during the Edo period" and "The import of Chinese sugar in the Nagasaki Junk trade and its impact" in this volume.

56 For example, the domestic production of sugar grew, and the amount of sugar on the Osaka market increased from 1804 onwards. Consequently, the price of imported sugar fell and the income of the *Nagasaki kaisho* decreased. Therefore the Shogunate limited the amount of domestic sugar on the Osaka market in 1808. However, people did not obey this limit and the price of imported sugar fell further. See Yamawaki, *Nagasaki no Tōjin bōeki*, 204–5.

57 Arano, "18seiki no Higashi-Asia to Nippon," 23.

If the Shogunate had sold the copper at a high price, they could have obtained foreign goods also only at a high price. In any case, foreign goods were expensive for the consumers in Japan.

6 The Copper Agency and Copper Merchant Houses in Osaka

Crude copper from all over Japan was gathered in Osaka where important parties such as the Copper Agency, refineries and copper merchant houses[58] were located. For the copper transport to Osaka, privately owned vessels were chartered. In the case of the Nanbu domain, the domain's officials in Osaka—in practice trading houses handling Nanbu products (*Nanbuton'ya* 南部問屋)—or copper merchant houses in Osaka chartered privately owned vessels from Osaka, Sakai, Hyōgo, the coastal area of Seto Inland Sea, or Shikoku Island through a *kaisendon'ya* 廻船問屋,[59] a kind of shipping agent.[60]

In Osaka, the copper merchant houses received the copper from the mines, forwarded it to the Copper Agency in Osaka,[61] received the payment from the Copper Agency, sent this to the managers of the mines, and obtained a commission.[62] In the case of the Akita domain, it was also a copper merchant house that received the crude copper in Osaka. Every time copper arrived in Osaka, officials of the Akita domain and the copper merchant house reported to the Copper Agency. The crude copper was directly forwarded to the copper refineries, where it was refined and the copper for export was cast in the form of bars (*saodō* 棹銅, copper bar). As soon as the copper bars were ready, the Akita officials and refineries reported the amount of copper bars produced to the Copper Agency. Then the Copper Agency chartered private ships through shipping agents specialized in the transportation of copper bars (*goyōdō kaisendon'ya* 御用銅廻船問屋)[63] on behalf of the *Nagasaki kaisho*, which

58 In case of the Besshi mine, the Sumitomo Family was in charge of mine management and at the same time the refinery and also played a role as copper merchant house.

59 Besides this, a *kaisendon'ya* 廻船問屋 went through the necessary procedure at a port, collected cargo for the ship, sold the goods carried by the ship, and took care of the crew. It was a combination of transport company and wholesaler. See *Kokushi daijiten*, s.v. "funadonya 船問屋".

60 Kudō, "Nanbu-han goyōdō-sen Tokujumaru no nansen shimatsuki," 81.

61 Before the Copper Agency was established, copper merchant houses passed the copper to copper refineries in Osaka.

62 It seems that the most of the copper merchant houses traded also other goods. Imai, "Kinsei Osaka no dō-kanren-gyōsha," 195, 198.

63 Iwasaki, "Nagasaki kaidō to sono kikō," 364.

was in charge of the copper transport from Osaka to Nagasaki.[64] The copper bars were loaded onto the chartered ships and then checked by officials of the Copper Agency. After that, the copper merchant house received the money for the copper. When the quota for an entire year was completely loaded onto a ship bound for Nagasaki, the Akita officials finally received a down-payment on the copper production of the following year.[65]

Besides in the copper transport, the copper merchant houses also played an important role in fundraising for the mines. The Akita domain, for example, was in financial difficulties like most of the other domains. The domain borrowed money from a copper merchant house not only for copper production and copper transport, but also to cover the costs of the domain's residence in Edo[66] as well as its running costs.[67] The copper merchant house lent the money on the security of the Copper Agency's down-payment for the following year's production. Since the official copper was monopolized by the Shogunate, it was a quite valuable security for borrowing money. The down-payment from the Copper Agency was put into the operating funds of the copper merchant house and invested in the market, not directly in the copper production. Thus, it was advantageous for the copper merchant houses when the official copper they held as security was evaluated lower than the actual market price. This means that the copper merchant houses were in favour of the low price of official copper. It is highly probable that this influenced the Shogunate's decision to keep the low copper price for such a long time.[68]

Historians Shimada Ryūto 島田竜登 and Sasaki Jun'nosuke 佐々木潤之介 argue that the domestic consumers were forced to buy copper at a high price and thus bore the financial burden of the 'irrational' copper pricing.[69] This is based on the analysis of the prices of Akita copper in 1773, which showed that the loss from sales of official copper was equivalent to the surplus from sales of

64 Before 1738 the refineries (*fukiya* 吹屋) in Osaka were in charge of copper transportation from Osaka to Nagasaki. See Iwasaki, "Nagasaki kaidō to sono kikō," 359.

65 Iwasaki, "Nagasaki kaidō to sono kikō," 364; *Daiisho*, 43–44; Imai, "Kinsei Osaka no dō-kanren-gyōsha," 197–98.

66 For *sankin-kōtai* 参勤交代 (alternate attendance) *daimyō* was stationed in Edo every other year. The expenses in Edo and the transportation to and from Edo were a big financial burden for *daimyō*.

67 Iwasaki, *Kinsei dōza no kenkyū*, 195.

68 See Iwasaki "The Akita domain and Osaka merchant houses at the time of the establishment of the Meiwa Copper Agency" in this volume.

69 Sasaki, "Dōzan no keiei to gijutsu," 184–85; Shimada, *The intra-Asian trade in Japanese copper by the Dutch East Company during the eighteenth century*, 57–59.

copper for domestic consumption.[70] As mentioned above, the Copper Agency bought the copper at two different prices: The price of copper for official purposes had been fixed low since the middle of the 18th century.[71] The price of copper for domestic consumption changed according to the situation at the time: the price the Second Copper Agency (1738–1746) paid changed depending on the market price.[72] After 1744 the copper for domestic consumption was freely traded.[73] The Third Copper Agency (1766–1868) paid the mines depending on their production costs.[74] The data which Shimada and Sasaki used for their analysis stems from the time of the Third Copper Agency.

The Third Copper Agency had a monopoly on the sale of crude copper. Therefore, a price manipulation as described by Shimada and Sasaki was possible. The Agency itself made good profit by selling copper for domestic consumption.[75] However, one should not ignore the fact that the Third Copper Agency was established in order to stabilize the market price as a reaction to its dramatic rise.[76] Since then, the copper market stayed stable for a long period thanks to the Copper Agency.[77] It is true that this Agency asked a high price from the consumers. At the same time, however, the consumers surely profited from the stable market price.

7 Copper Production

The volume of copper production declined in the early 18th century and stagnated afterwards simply because easily exploitable deposits near the surface became exhausted. Because the mines had to be driven deeper, more labour was required for excavating, drainage, and hauling out the ore, not to expand mining but just to maintain production at its former levels. Furthermore, charcoal, which was necessary for smelting, had to be procured from a distance since the neighbouring woods were exhausted. This caused additional costs

70 Sasaki, "Dōzan no keiei to gijutsu," 184–85; Shimada, *The intra-Asian trade in Japanese copper by the Dutch East Company during the eighteenth century*, 57–59.

71 Imai, "Kinsei Nihon no dō," 45.

72 Regarding the second Copper Agency see Imai, "Kinsei chūki no chiuridō ni tsuite," 61.

73 Imai, "Kinsei Nihon no dō," 40.

74 Ibid., 44.

75 Ibid., 41.

76 Ibid., 41.

77 Ibid., 43, 51–52.

and consequently the production costs rose.[78] In the Besshi 別子 mine, the production cost exceeded the selling price of official copper to the Copper Agency from the beginning of the 19th century.[79]

Under such circumstances, why did the copper mines continue production of official copper, although they made no profit by selling it? Did they really continue the production only because the Shogunate forced them to? Historian Nagazumi Yōko 永積洋子 tried to answer this question with the explanation that no one wondered about the low copper price at that time even though it was much lower than the market price simply because the production of official copper was seen as a kind of public service.[80] Historian Yamaguchi Keiji 山口啓二 also argued that the main purpose of the mines was to serve the Shogunate by producing copper and paying business tax to the Shogunate, but not to make a profit by copper production.[81] Indeed, the quota was seen as a kind of duty to the Shogunate.[82]

There was, however, another, more important motivation for the mines to engage in copper production. Firstly, copper production was heavily subsidized by the Shogunate. In case of the Besshi mine, the official copper price was determined at 139.48 *monme*/100 *kin* in 1751, and since then the price remained fixed until 1864. The Shogunate, however, allocated a subsidy called *teategin* 手当銀 to the mines. Under the name of *teategin*, the Shogunate actually paid more than the fixed price corresponding to the rising production costs of the time. Nevertheless, even with this subsidy the Besshi mine still could not fully cover production costs as shown in figure 4.3.[83]

Secondly, in case of disasters like flood or fire, the Shogunate lent a large amount of money to the mines for restoration or damage compensation.[84] However, it was seldom the case that this money was indeed spent for restoration work. Mostly, the money was invested in other realms. In the case of the Besshi mines of the Sumitomo Family, for example, the money went to the financial business of the Sumitomo in Edo and Osaka and was invested in the markets.[85] That means that the Sumitomo had the possibility to raise funds for their financial business under the name of copper mine restoration.

---

78 Yasukuni, "Kinsei Besshi dōzan no shūshi kōzō," 34; Sumitomo iinkai, *Sumitomo Besshi kōzanshi*, 98–100, 184.

79 Sumitomo iinkai, *Sumitomo Besshi kōzanshi*, 242.

80 Nagazumi, "Osaka dōza," 418–19.

81 Yamaguchi, "Kingin-zan no gijutsu to shakai," 160–61.

82 Iwasaki, "Kinzoku-shi kinsei Nihon-dō no seisan to ryūtsū", 29–43.

83 Sumitomo iinkai, *Sumitomo Besshi kōzanshi*, 241–43.

84 Yasukuni, "Besshi dōzan to Tenpō, Kaei-ki no keiei kiki," 76, Table 16.

85 Ibid., 44.

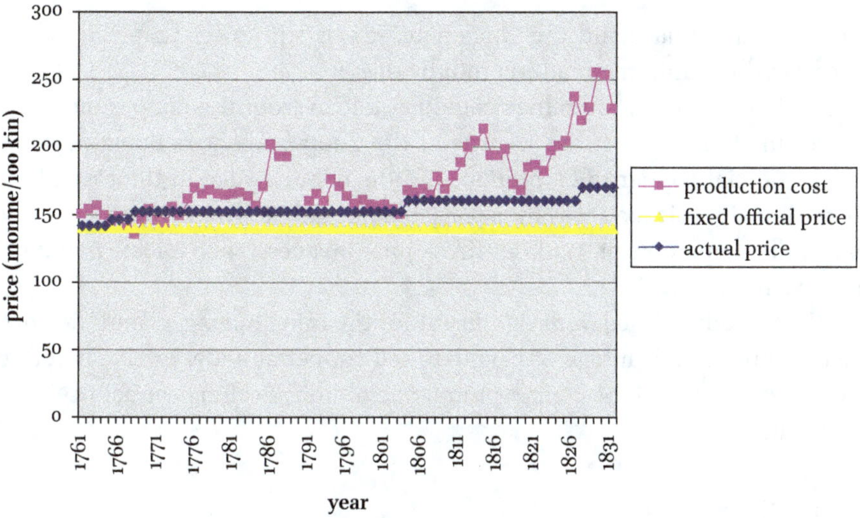

FIGURE 4.3 *The production costs, the fixed official copper price and the actual payment which the Besshi mine received from the Shogunate.*[86]

Thirdly, the mine managers benefited by selling daily necessities or lending money to the mine workers.[87] The mines were special areas which were isolated from outside society. There were guards stationed at the entrances to the mining area, and officials controlled all goods and people who entered and left the mine around the clock. A person who had no business with the mine was not allowed to enter the mining area, and the miners were not allowed to leave the area for a side job.[88] In such isolated areas, the mine managers were able to monopolize the sale of foods and other daily commodities and benefited from this.[89]

Fourthly, the mine managers profited from the rice purchasing system called *kaiukemai* 買請米 system. The Shogunate collected an annual tribute in rice which was/is the staple food of the Japanese. Since 1702, the Shogunate sold the rice which was collected as annual tribute in one of the *tenryō* 天領 (territory of the Shogunate) surrounding the Besshi mine, to the Sumitomo as *kaiukemai* at a lower price than the market price. This was also a kind of subsidy. In the mining area, the Sumitomo then sold this rice at a higher price

86 Ibid.

87 Yasukuni, "Kinsei Besshi dōzan no shūshi kōzō," 91.

88 Ishikawa, *Ikuno ginzan to ikuno daikan*, 196, 233.

89 Yasukuni, "Kinsei Besshi dōzan no shūshi kōzō," 10–11.

than what they had paid the Shogunate. It was still lower than the market price, yet the Sumitomo made a profit by this.[90]

Last but not least, the mines could get a loan from the copper merchant houses in Osaka against the security of the Copper Agency's down-payment for the official copper to be produced in the coming year. In the case of the Akita domain, this loan was indispensable not only to run its mines but also to run the domain itself. Without the copper business, such a loan would not have been available.[91]

All these advantages were significant for the mine managers and they outbalanced the disadvantage of having to sell copper at a low price. Therefore, the mines continued the copper production although selling copper itself was not profitable.

8 Conclusion: 'Reason and Order' of the Copper Trade

The above analysis shows that each party/stakeholder had some advantages by keeping the seemingly 'irrational' copper pricing.

Firstly, it is true that the mines suffered from the low copper price on the one hand. On the other hand, they received subsidies, loans and financial support for repair work in the event of accidents or disasters. The Sumitomo could invest this money in their financial business. The Akita domain could procure money from the copper merchant houses against the following year's copper production as a security. Therefore, the copper business was indispensable for the mines even though they could not make profit by selling official copper. Without the copper business, they had no chance to obtain this funding. Secondly, for the copper merchant houses who lent money against sales revenue of official copper as security, it was advantageous if the value of official copper was below the market price. Thirdly, the copper consumers in Japan profited from the stable copper market under this system. Fourthly, the low copper price was also welcome for the Japanese merchants who dealt in foreign goods. If the *Nagasaki kaisho* had offered the copper at a high "price" to foreign merchants, the *kaisho* would have received foreign goods at a correspondingly high "price" and consequently, the Japanese merchants would have had to pay more to the *Nagasaki kaisho* for these foreign goods. Then they would not have made so much profit by selling these goods. Fifthly, the *Nagasaki kaisho*, Nagasaki city and the Shogunate profited from the compensa-

90 Sumitomo iinkai, *Sumitomo Besshi kōzanshi*, 63, 67, 157, 159.

91 Iwasaki, *Kinsei dōza no kenkyū*, 195.

tion trade because the loss incurred by selling copper was well compensated by the margin obtained by selling import goods at high prices. Last but not least, the Chinese merchants also benefited from buying copper at a low "price" since the Chinese officials bought Japanese copper from them at a lower price than the market price.[92] The low copper price was thus favourable or at least acceptable for most of the parties involved.

Until now it has been assumed that the Shogunate strictly controlled the copper production, circulation, market and pricing through its absolute power. However, this view does not fit anymore to the result of the analysis presented in this paper. As Imai explained in her paper, the Shogunate's policy regarding copper was not always as coercive as usually assumed.[93] The fixed low price of official copper reflects not only the intentions of the Shogunate, but also the advantages of other stakeholders. This is why, the seemingly 'irrational' low price of copper remained for over hundred years.

---

92 Yamawaki, *Kinsei Nitchū bōeki-shi no kenkyū*, 74. Liu, "Shinnichi bōeki no yōdōshō ni tsuite," 112.

93 Further she argued that in principle the Shogunate reacted to the development of the copper market. See Imai, "Kinsei Nihon no dō," 58.

CHAPTER 5

Copper Administration Reform and Copper Imports from Japan in the Qianlong Reign of the Qing Dynasty

Liu Shiuh-Feng 劉序楓

1 Introduction

"Copper administration" refers to government policies to secure copper, which in imperial China was the most important raw material required for the minting of money. In the Qing Dynasty, both silver and copper money circulated. However, silver was primarily used for reference, for calculating the real market value of copper currency.

That is to say, in theory one silver *liang* 両 was equivalent to 1000 copper cash. But in times when copper currency was in short supply, its value would appreciate against the silver *liang* (i.e. one silver *liang* would be worth less than 1,000 copper cash). And of course the reverse happened when copper currency was debased.

In any case, the main point here is that securing a steady supply of copper for the minting of "official standard copper cash" (*zhiqian* 制錢) was vital in Imperial China because this was the currency commonly used in people's daily lives.

The first 100 or so years of the Qing Dynasty were, generally speaking, mainly years in which copper was in short supply, and its value appreciated against the silver *liang*. The main reason is that there was not enough official standard copper cash in circulation. The minting of money and the securing of necessary raw materials were problems that continually plagued the Qing Dynasty.

In the early years of the Qing Dynasty, domestic production of copper was inadequate. Besides using the old coins from the preceding dynasties and copper objects, the Qing Dynasty mainly relied on imported copper from Japan. Large-scale imports of Japanese copper were vital in stabilizing society and the economy in the early Qing.

However, in the 18th century, Japan began to limit its exports of copper because of the decline of copper production there and to meet its own needs. This move had a considerable impact on the Qing government, as evidenced by the frequent reforms the Qing adopted for its copper procurement system.

 | DOI 10.1163/9789004304512_006

The Yongzheng era government (1723–1735)—which had been rather passive about exploiting domestic copper sources—had no choice but to seek alternative sources of copper. The active exploitation of Yunnan copper mines was a result of these developments.

During the Qianlong reign (1736–1795), with the development of society and the economy and a sharp increase in population, the demand for copper currency rose considerably. It was risky to rely exclusively on imports of copper from Japan. Indeed, imported copper alone was no longer sufficient to meet domestic needs.

As a result, copper administration was reformed in the early years of the Qianlong reign. The metropolitan mint began to source its copper from Yunnan, rather than from Japan. Copper imports did not cease entirely, however. In order to simplify the copper administration, a new system was adopted in which designated official merchants or "quota merchants" (*eshang* 額商, a limited number of private merchants) were responsible for the procurement of copper and for supplying it to local monetary authorities. This system was maintained until the 1860s.

China continued to import copper throughout the Qing Dynasty. This paper will focus mainly on the period from late in the Kangxi reign (1662–1722) to late in the Qianlong reign, and especially on the Qianlong reign, for several reasons. The Qianlong era was the period of the highest economic and social development in the Qing Dynasty. It marked the zenith of the Dynasty before it went into decline. In the period to be examined here, the population grew, demand for currency was great, and the problem of a shortage of copper coinage was especially serious.

In this era, systems were established for the large-scale production of Yunnan copper, for changing the sourcing of copper used in the metropolitan mints, for the transport of copper to the capital and provinces, and for the purchasing of foreign copper by official or quota merchants. Although the amount of imported copper declined, because of its high purity it was able to compensate for shortfalls in Yunnan copper; it remained indispensable to provincial governments in coining money.

With the consolidation of an acquisition-and-management system, the import and transportation of Japanese copper came under government control, and there are thus detailed records in official documents from the period.[1]

1 The documentary sources used in this study include: (a) *Gongzhong dang Kangxichao zouzhe* 宮中檔康熙朝奏摺 [Secret Palace Memorials of the Kangxi Period], hereafter GDKZ; (b) *Gongzhong dang Yongzhengchao zouzhe* 宮中檔雍正朝奏摺 [Secret Palace Memorials of the Yongzheng Period], hereafter GDYZ; (c) *Gongzhong dang Qianlongchao zouzhe* 宮中

Currently available studies on the subject of Qing Dynasty copper administration focus mainly on problems related to Yunnan copper, such as production[2] and transport.[3] Few studies focus specifically on imported copper.[4] The main purpose of this paper will be to explore various issues pertaining to copper policy reforms in the Qianlong era, including the following: the background to those reforms, the role of imported copper in Qing Dynasty copper acquisition, and the amounts of copper imported.

2 Changes in the Copper Procurement System in the Early Qing

2.1 *Copper Procurement and the Minting of Official Standard Copper Currency (Zhiqian)*

In the Qing Dynasty, the minting of money was mainly the responsibility of the Baoquanju 寶泉局 and the Baoyuanju 寶源局, which were the metropolitan mints under the Board of Revenue and the Board of Works, respectively. These mints were founded after the Qing had seized the capital (Beijing) in 1644. During these earliest years of the Dynasty, the Shunzhi era (1644–1661),

檔乾隆朝奏摺 [Secret Palace Memorials of the Qianlong Period], hereafter GDQZ; (d) *Zhupi zouzhe caizheng lei* 硃批奏摺財政類 [Imperially Endorsed Memorials, Financial Matters], hereafter ZZCL; (e) *Ming-Qing dangan* 明清檔案 [Archives from the Ming-Qing], hereafter MQD; (f) *Guanyu Jiangning zhizao Cao jia dangan shiliao* 關於江寧織造曹家檔案史料 [Archival Materials Relating to the Nanjing Superintendents of Imperial Silk Manufacturing Cao Yin and his Sons], hereafter CJD.

2 Yan, *Qingdai Yunnan tongzheng kao*; Ch'üan, "Qingdai Yunnan tongkuang gongye"; Sun, "The Copper Industry of Yunnan: An Historical Sketch"; Sun, "Ch'ing Government and the Mineral Industries Before 1800"; Peng, "Qingdai qianqi Yunnan tongkuangye ji qi shengchan xingzhi de chutan"; Xu and Wu, *Zhongguo zibenzhuyi fazhan shi di yi juan: Zhongguo zibenzhuyi de mengya*, especially 477–521. Chen, "18 Shiji Zhongguo Yunnan de tong shengchan"; Chang, "Qingdai Yunnan de 'fangben shoutong' zhengce"; Qiu, "Qingdai Yunnan tongkuang de jingying"; Saeki, "Shindai Yōseichō ni okeru tsūka mondai"; Chiu, "Shiba shiji Yunnan tongcai shichangzhong de guanshang guanxi yu liyi guannian."

3 Nakajima, "Shinchō no dōsei ni okeru yōdō to tendō"; Sun, "The Transportation of Yunnan Copper to Peking in the Ch'ing Period"; Kawakatsu, "Shin Kenryūki Unnandō no keiun mondai"; Kawakatsu, "Shin Kenryū shonen Unnandō no Chōkō yuso to toshi Kanko"; Lan, "Qingdai Diantong jingyong luxian kaoshi"; Kim, "*Copper Transports* out of Yunnan, ca. 1750–1850: *Transport* Technologies, Natural Difficulties and Environmental Change in a Southwestern Highland Area."

4 Hall, "Notes on the Early Ch'ing Copper Trade with Japan"; Yamawaki, Chapter "Nisshin dōbōeki no shomondai" ("Some Problems Regarding the Copper Trade Between Japan and Qing Dynasty China") in *Kinsei Nitchū bōeki shi no kenkyū*.

the currency they produced was essential for paying official salaries, provisioning the military, and stabilizing society and the economy.[5]

In 1645, the government issued a proclamation to reward private citizens for going overseas to purchase copper.[6] At that time the new dynasty was not yet in control of the whole country, and anti-Qing resistance remained strong on the Southeast China coast. Under these circumstances, it is easy to understand the urgent need of the Qing Dynasty for copper. However, the public response to this proclamation should have been very limited, leaving the problem of copper supply unresolved.

That same year (1645), three internal customs station superintendents (*guanchai* 關差), including the one at Chongwenmen 崇文門 in the capital, were instructed to take charge of copper administration. In 1647, mints (*qianju* 錢局) were set up in various provinces. However, before the final suppression in 1681 of the "Rebellion of the Three Feudatories", Yunnan was not yet under the direct control of the Qing Dynasty. As a result, the government copper supply had to be purchased from Chinese merchants in other areas of China (accounting for the larger part), or imported from Japan (a smaller amount), and supplemented by the compulsory purchase of old coins or copper objects from the populace (see table 5.1). The shortage became so acute that in Fujian, for example, in 1655 the governor there petitioned to temporarily halt the minting of money because the local supply of old copper objects and discontinued coins was exhausted, and there was no copper for the mint.[7]

By 1657, with much privately-minted debased copper currency in circulation, copper coinage suffering devaluation, and a shortage of copper, all provincial mints stopped producing money, with only the metropolitan mints remaining in operation. In an effort to eliminate debased copper currency, the metropolitan mints increased the weight of a copper coin first from one *qian* (or mace) to one *qian* two *fen* (or candareens), then to one *qian* two *fen* and five *li*, and finally to one *qian* and four *fen*. At the same time old coinage was compulsorily purchased and then destroyed in order to stabilize the value of the currency.[8]

5 *Qingchao wenxian tongkao* 清朝文獻通考 [Encyclopedia of the Historical Records of the Qing Dynasty, also translated as Comprehensive Scrutiny of Documents: Qing Dynasty] (hereafter QWT), Vol. 13, "Qianbi Kao" ("Currency Policy") Shunzhi 1.

6 *Huangchao zhanggu huibian* 皇朝掌故彙編 [Collected Historical Facts and Records of Qing Dynasty], Neibian chapter, Vol. 19, "Qianfa yi" ["Monetary Law No. 1"].

7 MQD, Vol. 24, 136 jian (hereafter A24–136) Shunzhi 12/11/20 Yi Yonggui.

8 QWT, Vol. 13, Qianbi Yi (Currency No. 1) Shunzhi 8 and 14.

TABLE 5.1 *Systems for copper acquisition in the Qing Dynasty and amounts of copper involved*

Year	Method of acquisition (1)	Sources of financial capital (2)	Copper type (3)
1645	Customs Superintendants [CS] (Boaquanju [BQJ]) Board of Works officials (Baoyuanju [BYJ])	Internal customs taxes [CT] and other revenues	Market copper
1660	BQJ (same), CS, LZC, BOW officials (BYJ)	CT, Lu tax [LT]	same
1664	CS	same	same
1679	CS, Salt Superintendant	CT, LT, Salt tax [ST]	Market copper, co[…] objects
1684	CS	CT, LT	same
1699	CS, Imperial Household Department Merchant (Neiwufu Shangren [NFSR])	CT	Japan copper [JC], market copper
1703	NFSR, Private merchants	CT, ST	JC
1705	same	Internal CT, External CT, ST	JC, Yunnan copper
1709	CS, NFSR	same	same
1713	NFSR, Private merchants	same	same
1715	Eight Provinces individually	Taxes from provincial budgets	JC and YC, copper objects
1721	Jiangsu (for 5 provinces) and Zhejiang (for 3)	Land poll tax, CT	same
1724	Import of Japanese copper [JC] by Jiangsu, Zhejiang, Fujian, Guangdong	same	JC

ial quota (4) (jin)	Japanese exports (5)	Notes
d of Works, 1,800,000		Official acquisition by the Board of Works terminated in first year of Kangxi reign
1.190	249,860	
	1,993,100	Use of salt tax for copper acquisition terminated in 1681
2,307	2,675,100	Official copper quota in 1686: 3,642,758
1,160	5,236,435	
1,160	3,943,032	Increased purchase of 500,000 jin using salt tax
1,160	4,967,641	Additional acquisition using customs and salt taxes; Yunnan establishes government copper bureau
	5,170,521	5 of 14 internal customs superintendants purchase copper
ks) 2,520,000	3,909,199	Complete return to merchant purchasing, with additional acquisition by Board of Works and Board of Revenue
5,200	763,731	Implemented in following year; Termination of merchant copper purchasing
	2,696,000	Implemented in following year
,400	976,000	Implemented in following year;

TABLE 5.1 *Systems for copper acquisition (cont.)*

Year	Method of acquisition (1)	Sources of financial capital (2)	Copper type (3)
	Purchase of Yunnan copper [YC] by Hunan and Hubei		YC, copper objects
1730	JC for Jiangsu, Zhejiang, Anhui, Jiangxi, Fujian	same	JC
	YC for Guangdong, Hubei, and Hunan		YC, copper objects
1734	Halt purchase of YC by Guangdong, Hubei, and Hunan Minting in Yunnan with direct shipment to Capital	Provincial and local funds	JC and YC
1736	JC via Jiangsu and Zhejiang customs	Land poll tax, CT	JC
	Copper production by Yunnan		YC
1738	Total procurement from Yunnan	Board of Finance: 1 million liang of silver per year	YC
1739	Private merchants purchase JC for local mints	Funded by private merchants themselves	JC
	YC supplies capital mints	State-funded	YC
1745	State and private merchants buy JC for seven provincial mints	State-funded, self-funded	same
	YC for capital and various provinces	State-funded	

ial quota (4) (jin)	Japanese exports (5)	Notes
8,800	2,528,000	Japanese copper exports in Yongzheng 2 and 3 (export records from 1724 and 1725); Permission given for open-market selling of Yunnan copper, beginning in 1723
2,000	3,528,000	Implemented in following year;
3,200		Compulsory purchase of copper objects by local mints in 1729; 5 provinces buying Japanese copper could also buy yunnan copper beginning in 1731
	2,794,000	Implemented beginning in 1737
,000		
,000		
,000		One year halt in purchase of Japanese copper
rtain		Capital switches to use of Yunnan copper (including zheng, hao, and yu types);
,440 (capital city)		Beginning in 1741, half of private copper imports purchased by the state (Zhejiang, Jiangsu)
merchant ,000		State merchant reduced to 500,000 jin in 1750
,440 (capital city)		Beginning in 1744, 60% of private copper imports purchased by the state (Zhejiang, Jiangsu, and Jiangxi)

TABLE 5.1 *Systems for copper acquisition (cont.)*

Year	Method of acquisition (1)	Sources of financial capital (2)	Copper type (3)
1755	JC via state merchant and quota merchants	same	same
	YC for capital and various provinces		
1769	Same, continuing into future	same	same

Sources: Note: Source abbreviations based on abbreviations used in endnotes. For columns (1), (2), and (3): QWT (see note 5) Vol. 13–18; Da Qing huidain shili, Vol. 214–18. For column (4): Ibid.; also CJD GDYZ, GDQZ, ZZCL (see note 1 for all four of these sources); also *Suzhou fuzhi* (*Suzhou Prefectural Gazetteer*) Vol. 19 (1883 edition). For column (5): *Sen'oku sōkō*, Vol. 9; *Suijinroku* (*Kaishū zenshū*); Liu Shiuhfeng, "Kyōho nenkan no Tōsen bōeki to Nihon dō" (see note 64).

In 1675, permission was given for private enterprises to exploit copper and lead mines. In 1679, government agencies were ordered to send all scrap copper objects and brass cannon scrap to the Board of Revenue and the Board of Works to mint money. At the same time, a number of salt controllers—those in Lianghuai 兩淮 ("the two Huai," referring to the areas north and south of the Huai River), Liangzhe 兩浙 ("the two Zhe," referring to the eastern and western parts of Zhejiang province), Changlu 長蘆 (the Tianjin area), and Hedong 河東 (Shanxi and parts of Henan)—were instructed to use the silver acquired through the salt tax to undertake acquisition of copper.[9]

In sum, then, during the early Qing Dynasty, before the ban on maritime trade was lifted, the main sources of copper material for minting were still old currency and copper objects. The government continued to be plagued by a number of problems caused by the inadequate supply of copper, including the melting down of official currency to produce more coins, private minting of money, and the widespread use of coinage from former dynasties.[10]

9 QWT, Vol. 14, Qianbi Er [Currency No. 2] Kangxi 14, 18.

10 Cheng, *Mingmo Qingchu de yingui qianjian xianxiang yu xiangguan zhengzhi jingji sixiang*, 65.

al quota (4) (jin)	Japanese exports (5)	Notes
merchant oo te merchants ,000 than 9,000,000 onal total)	1,869,729	
rted 985,906; an (same)	1,280,565	State 505,906 jin; private merchants 480,000 jin (state purchase amount)

During the period of the Qing Dynasty when the "Ordinance of Evacuating from the Coast" (*qianjie ling* 遷界令) was in effect (1661–1683), anti-Qing forces, including the Zheng clan, the Jingnan 靖南 and Pingnan 平南 feudatories (in Fujian and Guangdong respectively), monopolized overseas trade. They imported silver and copper from Japan as capital, for making weapons, and minting money.[11]

In 1683, the Qing Dynasty succeeded in re-unifying the country, and the maritime trade ban was lifted, allowing citizens to engage in foreign trade. Looking just at Japan, the number of ships going there for trade grew rapidly, from 24 in 1684, to 85 in 1685, to 102 in 1686, and to 194 in 1688. (See appendix 1, which does not include those ships ordered to return to China because they were not given permission to trade.) From this point on, Japanese copper began to flow into China in large amounts, and this imported copper became the major source of material for the Qing Dynasty to mint money. Next, let us examine the various methods adopted by the Qing Dynasty for the acquisition of copper.

11 Liu, "Jūshichi jūhachi-seiki no Chūgoku to Higashi Ajia: Shinchō no kaigai bōeki seisaku o chūshin ni," 95–96.

2.2 *Changes in the Copper Procurement System*

As table 5.1 indicates, the Qing Dynasty frequently altered the copper procurement system, especially during the first 100 years of its rule. The main reason was simply the shortage of copper, the supply of which could not satisfy official quotas. Other major factors behind reforms in the copper procurement system were shortcomings specific to given systems in themselves, and also the problem of corruption.

The copper procurement systems described in table 5.1 applied mainly to the Baoquanju and Baoyuanju mints in the capital. As for provincial mints, these operated only sporadically in the early Qing Dynasty (because of the lack of copper), and they did not begin to operate on a stable basis until after the reforms of the early Qianlong reign.

Let us first look at the systems used in the early Qing Dynasty. During the Shunzhi and Kangxi reign periods, customs station superintendents were in charge of copper acquisition. Beginning in 1699, this task was turned over to "Imperial Household Department merchants" (*neiwufu shangren* 內務府商人). Detailed studies have been made of these systems,[12] so here I will only highlight the major points.

The "customs-station-superintendent-copper-acquisition system" began in 1645. At first, four internal customs station superintendents—Chongwenmen (in the capital), Tianjin, Linqing, and Huai'an—were required to convert 10,000 *liang* of silver from customs revenues into copper, and turn this copper over to the metropolitan mints. These numbers were increased over time to include ten internal customs station superintendents and a total of 160,000 *liang* of silver.

In 1660, the Baoyuanju altered the system to include copper procurement by both customs station superintendents and the so-called "reed administration superintendents" (*luzheng chai* 蘆政差). However, the official purchase price (6.5 *liang* of silver per 100 *jin* of copper) was far below the market price (see table 5.2), and also well below the price of copper from Japan (which was exported at prices ranging from 10.4 to 13.5 *liang* per 100 *jin* in the years 1697–1712).[13] In order to make up the difference (that is, to increase the supply of silver to pay for the copper), customs station superintendents resorted to the arbitrary imposition of customs taxes, which disrupted commerce and caused

12 Kōsaka, "Shindai zenki no kansa bendōsei oyobi shōnin bendōsei ni tsuite"; Dunstan, "Safely Supping with the Devil: The Qing State and Its Merchant Suppliers of Copper."

13 Yamawaki, Chapter "Nisshin dōbōeki no shomondai" in *Kinsei Nitchū bōeki shi no kenkyū*, 74, 137.

TABLE 5.2 *Prices of Japanese and Hankou copper from the Kangxi to the Xianfeng Reign periods (unit: silver* liang *for 100* jin *of copper)*

Year	Imported copper		(3) Market price in Hankou	Notes
	(1) Official price	(2) Market price		
1664	6.5	10–14 ①		Market prices in various provinces, 1670
1684	"	14–15 ②		Copper market price (non-imported copper)
1686	10	16–17 ③		Same as Hubei/Hunan market price for 1688
1716	12.5			
1719	14.5			
1727	"	18–19 ④		Jiangsu-Zhejiang market price in 1723
1740	17.5	19.8–20 ⑤		Jiangsu-Zhejiang market price
1745	Official merchants 13, 14 Private merchants 17.5			
1746	14.5 ①*		20.68 ①	* Price of imported copper in Hubei
1749	Official merchants 13, 14 Private merchants 17.5	20–22 ⑥	19 ②	Zhejiang market price
1763	"	19~ ⑦	18.95 ③	Fujian market price
			19→17.5 ④	
1770	"	25.079 ⑧		Suzhou market price
1772	"		17.5 ⑤	
1784	Official merchants 13, 14			

TABLE 5.2 *Prices of Japanese and Hankou copper* (cont.)

Year	Imported copper		(3) Market price in Hankou	Notes
	(1) Official price	(2) Market price		
	Private merchants 15.3 ②			
1797	Official merchants 13.593 Private merchants 15.3 ③	24.79 ⑨		Suzhou market price
1808	Official merchants 12 Private merchants 15.3 ③			
1835	Official merchants 13.593 Private merchants 15.3 ③	24 ⑩		1856 Shanghai market price

Sources: For column (1): QWT (see note 5), Vol. 13–17; *Qianlong huidian zeli* (see note 57), Vol. 44; *Da Qing huidian shili* (see note 80), Vol. 214–218
①. MQD (note 1), A.144–145
②. GDQZ (note 1), Vol. 60, 75
③. *Lin Wenzhonggong zhengshu jiaji* [Political Correspondence of Lin Tse-hsu], "Jiangsu zougao"
For column (2):
①. *Huangchao jingshi wenbian* (note 52), Vol. 53; Yao Wenran.
②. *Qing shilu*, Kangxi period, Vol. 116, 19b
③. *Qingshi liezhuan* (note 47), Vol. 9, "Song Luo"
④. GDYZ (note 1), Vol. 9, 5
⑤. MQD (note 1) A93–63
⑥. ZZCL (note 1), 1239–014
⑦. GDQZ (note 1), Vol. 20, 293
⑧. *Suzhou fuzhi* [Suzhou Prefectural Gazetteer] Vol. 19 (1883 edition): 15a
⑨. *Fujian shengli* [Laws and Regulations of Fujian Province], Taiwan wenxian congkan, No. 199 (1964): 357
⑩. "Cheng Jiatang shizai daomai gongju tongjin jielüe" ["Description of the Illegal Sale of Quota Copper by Cheng Jiatang"]. Manuscript, kept at the Nagasaki Museum of History and Culture.
For column (3):
①. ZZCL, 1237–029, MQD A144–115
②. MQD A153–59
③. GDQZ, Vol. 17, 463
④. QWT, Vol. 17, *Qinding Da Qing huidian shili*, Vol. 218
⑤. ZZCL 1284–024

disaffection among the public. In any case, the system was not very effective in its goal of ensuring a steady supply of copper.

The Kangxi emperor eventually recognized the seriousness of this problem. In 1686, the official purchase price for copper was raised to 10 *liang* of silver per 100 *jin* 斤 of copper. However, this was still well below the market price. Moreover, many officials deliberately obstructed the copper acquisition process for the following reason: Customs officials were supposed to change each year. But there was a rule that if an official's annual quota was not fulfilled, he could remain in office until he made up the difference. Thus, for personal gain, many officials deliberately procrastinated in the procurement process.[14] Given these problems, in 1699, another new system was introduced in which silver from customs station superintendents was turned over to Imperial Household Department merchants for copper purchases.

Under the "imperial-household-department-merchants system," these merchants received silver from customs station superintendents, purchased the copper themselves, and delivered their quota to the metropolitan mints. Beginning in 1700, the entire quota for copper—3,581,160 *jin* from fourteen internal customs station superintendents—was turned over to nine merchants, including Wang Gangming 王綱明, Zhang Dingchen 張鼎臣, and the Cao Yin 曹寅 brothers.[15]

In 1703, because the weight of official standard copper coins (*zhiqian*) was increased, even more copper was required, and salt-tax silver from Changlu, Shandong, Lianghuai, Hedong, and Liangzhe was also allocated for this purpose. In 1705 and 1713, salt-tax and maritime-customs silver from Fujian and Guangdong was added to the funds earmarked for copper procurement. At this time all copper procurement was still commissioned to Imperial Household Department merchants.[16]

For 100 *jin* of copper, the government provided the merchants with ten *liang* of silver, plus five *liang* for transport. Merchants in fact used only seven *liang* to purchase the copper, and another three *liang* went on transport and miscellaneous expenses. This total of ten *liang* was still below the price of both Japanese exports and the domestic market. Of the surplus five *liang*, 1.1 was given to the customs authorities as "travel expenses" (*panchan yin* 盤纏銀). The other 3.9 was considered "saved silver" (*jiesheng yin* 節省銀) and was deposited in the treasury.[17]

---

14 GDYZ, Vol. 24, 762, Yongzheng 13/6/2, Li Wei.

15 CJD, 15–20, Kangxi 40/5/23, Neiwufu (Imperial Household Department).

16 QWT, Vol. 14, Kangxi 42, 44, and 52.

17 CJD, 71–72, Kangxi 48/6/4, Neiwufu.

As there was virtually no direct profit for the merchants who engaged in copper acquisition, they had to rely on profits from other products; or they might benefit from interest on silver given to them up front for eventual copper purchases, or from investing the silver in other activities. The most important benefit to Imperial Household Department merchants was that, by turning over "saved silver" or "travel expenses" to the higher authorities, they could establish excellent relationships in the imperial household or with the customs authorities, and thereby enjoy special privileges.[18] Thus, even if the allocation of silver for copper acquisition was low, the merchants did not necessarily lose out.

Eventually, however, mismanagement by Wang Gangming and other merchants led to accumulated losses of over two million *liang* of silver, and the merchants could not fulfil their quotas for copper delivery. After a petition by Zhao Shenqiao 趙申喬, president of the Board of Revenue,[19] the system was changed yet again. Beginning in 1716, eight provinces (Jiangsu, Anhui, Zhejiang, Jiangxi, Fujian, Hubei, Hunan, and Guangdong) undertook copper procurement individually.

Under the "eight-provinces copper-acquisition system," the governor of a province assigned officials to appropriate government silver and go to ports in Zhejiang and Jiangsu, where they would seek out merchants to acquire copper. Each province was assigned a quota of 554,400 *jin* of copper, for a total of 4,435,200 *jin*. The official purchase price was raised to 12.5 *liang* per 100 *jin*, plus an additional three *liang* for transportation costs.[20]

It was around this time, in 1715, that Japan instituted the Shōtoku Regulations 正徳新例. These set limits on the number of Chinese ships that could come to trade, and also on the amount of copper exported. Japan began a system of issuing *shinpai* 信牌 (tallies or permits for trading at Nagasaki). Those without permits were not allowed to engage in trade. Merchants seeking to secure permits got into legal disputes with one another, and Jiangsu and Zhejiang customs confiscated all the *shinpai*. As a result, in 1715 and 1716 merchant ships were unable to go to Japan[21] and the acquisition of Japanese copper was interrupted for a time. The metropolitan mints had no choice but to buy as many

18 Kōsaka, "Shindai zenki no kansa bendōsei oyobi shōnin bendōsei ni tsuite," 141–45. Also see Wang, "Kangxi cao houqi de tongzheng gaige yu Neiwufu guanshang," 61–72.

19 *Kangxi qijuzhu*, ce 3: p. 2173.

20 QWT, Vol. 14, Kangxi 54.

21 Liu, "Shindai zenki no Fukken shōnin to Nagasaki bōeki," 140–43.

old copper objects as they could as a temporary measure. In 1718, the official price for copper was again raised, to 14.5 *liang* of silver.[22]

Another problem was that officials from the six provinces other than Zhejiang and Jiangsu were not aware of conditions in Jiangsu and Zhejiang, and were not familiar with the maritime merchants. They were unable to negotiate prices effectively with the merchants, thereby wasting public funds. And still the required amounts of copper could not be delivered on time. Hence, in 1722 the system was revised. While each of the eight provinces was still responsible for the same quota of copper as before, under the new system Jiangsu would act as agent for five provinces, and Zhejiang for three.

However, regardless of the type of copper procurement system adopted by the Qing Dynasty, the problem remained that Japan prohibited the export of large quantities of precious metals. In 1685, it set limits on the number of Chinese merchant ships, trade volume, and copper exports.[23] As mentioned earlier, the implementation of the Shōtoku Regulations in 1715 further limited the number of Chinese trading ships, to thirty per year, and the export of copper to three million *jin*. Thereafter it became extremely difficult for merchants to fulfil the government quota of 4.43 million *jin* of copper from abroad. This problem is mentioned in numerous official memorials[24] with the Japanese restrictions on exports remaining the key sticking point.

By 1725, however, the production of copper in Yunnan had increased considerably. Hunan and Hubei were instructed to begin acquiring their copper quotas from Yunnan at that time, and Guangdong was given similar instructions in 1731. This left five provinces buying from abroad and three from Yunnan. However, those provinces relying on imported copper were allowed to compensate for any shortage by also purchasing copper from Yunnan. Yunnan was producing growing amounts of copper—in the range of two to four million *jin* for 1726–1727.[25]

Another major reason why the Qing had continual problems acquiring copper and consequently repeatedly changed the rules for copper acquisition is that the official purchase price was too low, far below the market price. As table 5.2 indicates, during the reigns Kangxi and Qianlong, the market price for imported copper was much higher than the amount offered by the government.

22 QWT Vol. 14, Kangxi 55, 56, and 57.

23 See Nakamura, *Kinsei Nagasaki bōeki shi no kenkyū*, 372–75.

24 GDKZ Vol. 7, 547, Kangxi 58/6/11 Cao Fu; QWT, Vol. 15, Yongzheng 2, quoting Ho Tianpei.

25 GDYZ, Vol. 8, 353. Yongzheng 5/6/17, Yang Mingshi and Vol. 10, 473, Yongzheng 6/5/21, E Ertai.

There were frequent reports of merchants profiting by selling government copper to private citizens.

Moreover, with the price of Japanese copper being much higher than the official price, merchants could clearly see that there was virtually no profit to be made from it. Those still willing to accept responsibility for the acquisition of copper on behalf of the state did so generally in the expectation that the budget for purchasing the copper, which was received up front before they sailed for Japan, could be diverted to other profitable lines before being used for purchasing copper.[26]

On top of the already low official purchase price, officials directly in charge of copper procurement would often lower the price further in order to save money.[27] As a result, few ordinary merchants were willing to take responsibility for purchasing copper for the state. As the governor of Fujian Lu Zhao 盧棹 wrote in 1735: "With regard to merchants accepting responsibility for government copper, there is virtually no profit, and moreover they are subject to being 'squeezed' by the Japanese side. Thus, most wealthy merchants are unwilling to take the job."[28] Previous studies have all suggested that import copper merchants may have made huge profits, but, at least as far as the procurement of copper for the state is concerned, this appears not to have been the case.

Here we may end our discussion of the copper procurement systems during the early Qing Dynasty—and the reasons for the many changes made. Let us summarize the systems from the Shunzhi to Yongzheng reigns as follows: In order to acquire copper for the metropolitan mints, designated provinces (or customs station superintendents, or Imperial Household Department merchants) were given responsibility for this task. Officials directly in charge of copper procurement went to the coast of Jiangsu or Zhejiang to recruit merchants. First, silver from government funds, customs taxes, or salt tax was turned over to the merchants. These were allowed to load up cargo and sail to Japan to trade. In Japan they then had to use what they earned to purchase copper on behalf of the state and bring it back to China. They had to deliver the stipulated quota to customs officials in Jiangsu or Zhejiang or to copper-acquisition officials from other provinces who were waiting for them in Jiangsu or Zhejiang. Each of these officials would then deliver its share to the capital.

26 See GDYZ, Vol. 9, 5. Yongzheng 5/9/19, Li Wei and Vol. 24, 562, Yongzheng 13 leap month 4/20, Lu Zhao.

27 Ibid. Vol. 1, 686, Yongzheng 1/9/9 and Vol. 4, 113, Yongzheng 3/4/3, He Tianpei.

28 Lu Zhao op. cit. (note 26).

3 Reform of Copper Administration under the Early Qianlong Reign

3.1 *Development of Yunnan Copper Production*

It has already been mentioned in the previous section that, after Japanese restrictions on the export of copper led to a shortfall of copper in China, the Qing government was forced to devote more effort to the exploitation of Yunnan copper mines. Around 1716, production there was only just over 600,000 *jin*. It exceeded one million *jin* in 1723 and rose to 4.85 million *jin* in 1734. In 1735, the year in which the Qianlong emperor ascended to the throne, production suddenly boomed to more than 7.59 million *jin*.[29] Thus, within a twenty-year period, production increased by a factor of more than 10.

In fact, as early as 1682, after the suppression of the rebellion of Wu Sangui 吳三桂, Cai Yurong 蔡毓榮, governor-general of the Yunnan-Guizhou area, had already proposed "minting copper coins intensively" and "developing mine reserves" as financial strategies for paying the costs of maintaining the military garrison there.[30] This suggestion was accepted by the Kangxi emperor, and Yunnan's copper mines were increasingly exploited thereafter. In 1705, a state-run copper purchasing bureau was opened in the Yunnan provincial capital. It collected the required quotas from each copper mining operation and also compulsorily purchased any copper in excess of the quotas, which was then sold to merchants to provide to the capital for minting money.[31]

However, production was limited in this early period and remained low throughout Yongzheng's reign. The Yongzheng emperor had a passive attitude toward developing Yunnan copper mines, fearing that it would create concentrations of workers, which in turn would adversely affect social stability, lead to rebellion, or cause food shortages.[32] Moreover, the frequent rebellion of the various Miao people prior to 1727 hampered the development of mining.

In 1726 and 1727, E Ertai 鄂爾泰 successively subdued the various Miao communities of northeast Yunnan (including the Dongchuan, Wumeng, and Zhenxiong Miao). Local hereditary chieftainships were replaced with circulating (central government appointed) officials (*gaitu guiliu* 改土歸流). Han

---

29 GDYZ, Vol. 2, 189, Gao Qizhuo and others; 529 Jin Shiyang; ibid. Vol. 25, 44, Zhang Yunsui. Also, MQD A76–47, Zhang Tingyu and others.

30 Cai, "Chou Dian shi shu," 17–23.

31 QWT, Vol. 14, Kangxi 44.

32 GDYZ, Vol. 12, 33, memorial of Guangdong provincial financial commissioner Wang Shijun, rescripted by the Emperor. See also Saeki Tomi, op. cit. (note 2): 533–37.

Chinese were brought in to begin farming the land and developing copper mines. It was only then that copper production began to steadily increase.[33]

Nevertheless, full-scale exploitation did not begin until after the Qianlong emperor ascended to the throne. Historical documents do not clearly indicate the reasons for the rapid expansion of production from the late Yongzheng to the early Qianlong periods. However, it can be surmised that the Qianlong emperor ordered that production be increased, that new mines be opened, and that management be tightened up, in order to supply copper for the minting of money. In addition, Yunnan-Guizhou governor-general Zhang Yunsui petitioned to raise the official price for copper, and compulsorily purchased copper from the mines, contributing to the sudden increase in production.[34]

By the late Qianlong period, annual production in Yunnan had reached about ten million *jin* per year.[35] This created a solid financial foundation for reforms under which Yunnan copper replaced copper imported from Japan.

3.2 *The Shift in Copper Sourcing for the Metropolitan Mints*

After Japan instituted the Shōtoku Regulations in 1715, reducing Chinese imports of Japanese copper, the sources of copper for the metropolitan mints changed as the proportion coming from Yunnan increased. In the early Yongzheng period, the state halted the compulsory purchase of surplus copper (i.e. copper in excess of the quotas for each mining operation), and private merchants were permitted to sell this surplus copper themselves. (The compulsory purchase of surplus copper had greatly reduced incentives to produce copper over and above quotas.) In 1725, various mints in Yunnan were authorized to mint official standard copper cash (*zhiqian*), which was allowed to circulate directly to other provinces and did not need to go through Beijing.[36]

In 1731, the copper procurement system was again changed, to the aforementioned "eight provinces" system, with some provinces buying from Japan and others from Yunnan. But it remained impossible to fulfil the required quota for imported copper.

In 1722 and 1723 new rules were instituted to set time limits for the acquisition of copper and for making up shortfalls in previous copper quotas. If officials responsible for the acquisition of copper and its delivery to the capital failed to deliver on time, and failed to bring their accounts up to date within a year, they

33 Huang, "E Ertai yu Yun-Gui Miaojiang de jingying, 1726–1731," 19–45.

34 See Yang, "Zhang Yunsui yu Qingdai qianqi Yunnan shehui jingji de fazhan," 46.

35 See Yan, *Qingdai Yunnan tongzheng kao*, 81–84; Qiu "Qingdai Yunnan tongkuang de jingying," 82–83; Xu and Wu, *Zhongguo zibenzhuyi fazhan shi*, 491–93.

36 QWT, Vol. 15, Yongzheng 1 and 3.

and their superiors would be punished and ordered to pay compensation.[37] But these rules did not have much impact, and from the late Kangxi through the early Yongzheng reigns, although 75 officials were responsible under the rules, the shortfall in official copper remained, exceeding 3.8 million *jin*. At that time, "most copper procurement officials were ruined, and no copper purchasing merchants did other than lose their capital."[38] The losses in government funds accumulated steadily, and it became necessary to take action to fundamentally alter the system and put it on a sound footing.

Therefore, when the Qianlong emperor ascended to the throne, he ordered all ministries and governors to offer advice on reforming copper administration and developing copper resources. A whole range of arguments were put forward, some supporting the continued purchase of Japanese copper, others calling for an end to the importation of copper and trading with Japan; yet others called for the exclusive use of Yunnan cooper, and others for a combination of Yunnan copper and imported copper.[39]

Among the advice offered was that of Jiangsu governor Gu Zong 顧琮 in 1736. He put forward five proposals.[40] The main thrust of his advice was to use Yunnan, Sichuan, and Japanese copper concurrently, but to reduce import quotas. He also advised lifting the ban on copper trading that had been in place since the Yongzheng reign and allowing copper to be traded freely. Merchants could thus, besides using official copper currency, freely buy and sell surplus copper. This would give them more incentive and stabilize the price of copper.

Gu's proposals were endorsed in principle by Board of Revenue president Zhang Tingyu 張廷玉 and others. After the weight of copper cash had been reduced from one *qian* four *fen* to one *qian* two *fen* in 1734, the quota for copper procurement could be lowered from 4.43 million *jin* to 4 million *jin*. Japanese copper and Yunnan copper were each to cover half this quota (i.e. two million *jin*). In turn, of the imported copper quota, Jiangsu and Zhejiang customs were each charged with acquiring one half (i.e. one million *jin* each). Yunnan and Japanese copper were to be concurrently acquired beginning in 1738.

Meanwhile, merchants were permitted to sell surplus copper, in order to stabilize the price.[41] Later, a number of individuals, including Jiangsu provincial

37 Ibid. Vol. 14, Kangxi 61, and Vol. 15, Yongzheng 1.

38 ZZCL, document no. 1226–027, Qianlong 1/1/17, Li Fu.

39 See *Zhupi zouzhe caizheng lei mulu* 硃批奏摺財政類目錄 (Imperially Endorsed Memorials, Financial Matters, Table of Contents) Vol. 4, 714ff.

40 ZZCL, 1226–030, Qianlong 1/1/25, Gu Zong.

41 Ibid. 1226–032 Zhang Tingyu and others, response to memorial. MQD A67–79. QWT Vol. 16 Qianlong 1.

finance commissioner Zhang Qu 張渠 and Grand Secretary Ji Zengyun 嵇曾筠, advised halting the acquisition of Japanese copper for 1738; they wanted to first resolve problems in the acquisition process and also clear backlogged debts and shortfalls. In addition, trading licenses were to be transferred to those merchants willing to sell their own copper to the state. Rather than give copper-purchasing funds (i.e. silver) to the merchants up front, the government would wait for the copper trading ships to return to China, after which customs officials would purchase a stipulated amount of the copper. Moreover, rules were established governing rewards and punishments for officials in charge of copper procurement; the "saved silver" system was abolished and the full amount of transport costs paid.[42]

The reforms of 1736 had a strong impact on the number of trading ships going to Japan in 1737 and 1738. As appendix 2 indicates, only five ships went in each of those two years. In 1737, because the production of Yunnan copper was so high—production (both high-grade and low-grade) totalled more than 7.5 million *jin*—a surplus of over three million *jin* remained after deliveries to the capital and to Yunnan for minting money as well as purchases by the various provinces.

Yunnan governor-general Yin Jishan 尹繼善 even petitioned that Jiangsu and Zhejiang shift their copper purchases for 1738 from Japan to Yunnan.[43] This proposal was opposed by Li Wei 李衛, the governor-general of Zhili and also a former governor of Zhejiang, who was in charge of overseeing copper acquisition. He thought it would be inefficient, given the enormous distance between those two coastal provinces and Yunnan, to dispatch officials to acquire copper and transport it all the way back. By having Yunnan ship its copper directly to Beijing, errors could be avoided.[44]

In the end, the "nine ministers" (*jiuqing* 九卿, the highest-ranking officials) decided that, beginning in 1739, Zhejiang and Jiangsu would acquire two million *jin* of Yunnan copper which would be shipped directly to the capital by Yunnan.[45] This decision meant that Beijing's copper requirement of four million *jin* would be met entirely by the province of Yunnan. Meanwhile, Zhejiang and Jiangsu could seek out merchants with their own financial resources and dispatch these to buy Japanese copper to supply local mints.

42 ZZCL 1227–011, Qianlong 1/5/24, Zhang Qu; 1227–005, Qianlong 1/10/13, Ji Zengyun. QWT Vol. 16 Qianlong 1.

43 ZZCL 1227–011, Qianlong 2/5/27, Yin Jishan; 1227–013, Qianlong 2/7/10, Zhuang Qin Wang Prince Yunlu; QWT Vol. 16 Qianlong 2.

44 ZZCL 1227–027, Qianlong 3/2/16, Li Wei; QWT Vol. 16 Qianlong 3.

45 ZZCL 1228–001, Qianlong 3/2/25, Hai Wang and others; QWT Vol. 16 Qianlong 3.

Yunnan copper production peaked after 1736. Besides opening new mines, the authorities also built and improved transport routes to ship the copper to the capital. In 1738, Yunnan Governor Zhang Yunsui petitioned on fifteen items connected with the transport of Yunnan copper to Beijing. After review by the Grand Secretary, the "Yunnan Copper Transport Regulations" were established. These set out rules for time limits on transport; for amounts of copper (including so-called *zheng* 正, *yu* 餘, and *hao* 耗 copper, to be explained below); for transport routes and costs; for the handling of incidents en route; and for the monitoring, rewarding, and punishment of officials.[46]

The same year—with strong impetus provided by Zhang Yunsui—roads were built or improved and the Jinsha River waterway was opened for transport as far as Luzhou in Sichuan.[47] From there copper shipments went by way of the Yangzi River to Hankou and thence to Yizheng in Jiangsu, from where they took the Grand Canal to Tongzhou and were finally delivered to Beijing. Thus began the large-scale transport of Yunnan copper to the capital.[48]

By 1749, a petition was made to set time limits for copper transport, owing to a number of problems involving copper and lead, delays by local authorities, and theft by local officials who would then sell the copper themselves. A limit of eleven months was set for the shipment of copper from Luzhou to Tongzhou. Provincial authorities were ordered to periodically report such information as the starting dates of shipments, dates of entrance into or departure from a given province, whether or not there were any incidents, and detailed explanations of delays caused by storms or freezing of rivers.[49]

Besides the annual quota of four million *jin* of "regular" (*zheng*) copper for the capital, it was also stipulated that an additional 1,704,000 *jin* be delivered to Beijing, making 5,704,000 *jin* overall. However, because the purity level was only 95 percent, a further eight *jin* of copper, known as "compensatory" (*hao*)

---

46 ZZCL 1228–008, Zhang Yunsui, Qianlong 3/5/30; 1228–015, Qianlong 3/7/9, E Ertai; QWT Vol. 16 Qianlong 3.

47 *Qianlong chao shangyu dang* 乾隆朝上諭檔 [Collected Imperial Orders] Vol. 1, 611–12, 685–88; ZZCL 1228–016, 1229–009, 1233–004, 1233–013, 1235–009, 1236–006, 1237–022, 1239–002, 1240–013, from Qianlong 3 to 14, are all related to the shipment of Yunnan copper to Beijing. Also see "Zhang Yunsui zhuan" ["A Biography of ZhangYunsui"] in *Qingshi liezhuan* [Qing Historical Biographies] Vol. 16, ce 4, 1183. Yang, "Zhang Yunsui yu Qingdai qianqi Yunnan shehui jingji de fazhan." Also Chen, "Qingchu Jinshajiang shuidao kaijun gaikuang," 200–5.

48 See Kawakatsu, "Shin Kenryūki Unnandō no keiun mondai"; Kawakatsu "Shin Kenryū shonen Unnandō no Chōkō yuso to toshi Kanko."

49 *Qing shilu* 清實錄 [Veritable Records of the Great Qing] Vol. 341, Qianlong 14/5/27; QWT Vol. 17, Qianlong 14.

copper had to be added for each hundred *jin*. Finally, to compensate for losses in shipping and other expenses, an additional three *jin* of so-called "extra" (*yu*) copper had to be supplied. As a result, the amount of copper transported to the capital each year was set at 6,331,440 *jin*.

Besides the capital, various provinces purchased roughly two to three million *jin*, and Yunnan needed about two million *jin* for its own minting needs.[50] In sum, it was necessary to produce about ten million *jin* annually in order to fully meet the nation's minting needs.

From 1737 on, the production of Yunnan copper maintained a level of roughly ten million *jin*; at its peak, the production reached more than thirteen million *jin*.[51] In 1775, Yunnan provincial finance commissioner Wang Taiyue 王太岳 was able to write in the "On Copper Administration": "According to my investigation, because there was a surplus in Yunnan of 3.74 million *jin* in the second year of the Qianlong reign, it was possible to discontinue purchasing Japanese copper."[52]

Thus, although initially reductions in Japanese copper imports had a negative effect on the of minting of official standard copper coins (*zhiqian*), this was the spark that led to greatly increased production of Yunnan copper until such a point that Yunnan copper could completely replace imported copper to meet all the capital's needs for copper for minting. The copper-currency-shortage problem, which existed throughout the early Qianlong reign, steadily eased after 1780 because there was a sufficient supply of copper, and various localities set up their own mints one after another and produced continually increasing amounts of currency.

3.3 *Copper Acquisition by State Merchants and Private "Quota" Merchants*

As discussed earlier, from 1739 onward the metropolitan mints of the Ministries of Revenue and works were able to switch over entirely to using Yunnan copper for minting. However, imports of copper continued where there was not enough Yunnan copper. That year, self-financed merchants were allowed to buy goods to take to Japan, to trade these there and buy copper. The trading licenses of merchants who owed the government money were confiscated and given to new merchants to trade abroad. What was different from the past is

50 Yan, *Qingdai Yunnan tongzheng kao*, 81–88; Xu and Wu, *Zhongguo zibenzhuyi fazhan shi*, 490.

51 See footnote 35. However, Yan Zhongping's estimate is on the high side, and is not as reliable as that from the documentary evidence used by Qiu Lijuan.

52 Wang, "Tongzheng yi," 25.

that these merchants had to have their own trading capital. After their return to China, their copper was bought by Jiangsu and Zhejiang.[53]

However, in 1737 and 1738, merchants did not respond very positively to this new policy. The main problems were a lack of capital and the low profits to be earned by acquiring copper.

In 1737, Jiangsu Governor Zhang Qu submitted a request to halt official compulsory purchases of merchant-acquired copper, asking instead that it be allowed to be traded freely in order to stabilize the price. In the past, even though the official price of 14.5 *liang* of silver per 100 *jin* of copper was far below the market price, merchants could at least first receive the purchase funds from the government and make up for the lack of profit from buying and selling copper by collecting interest or profiting from trade in goods for which the government funds served as the initial capital. Under the new system, in which merchants had to supply even the initial capital, why would any merchant be willing to use his own profits to buy copper at a high price only to be forced to accept the much lower compulsory official price when returning to China?[54]

In 1740, as copper currency became scarce in Jiangsu, the central government was called upon to allow minting to begin. However, because the copper supply was inadequate, Zhang Qu petitioned for the official purchase price of 14.5 to be raised by the 3 *liang* originally allowed for transport and other expenses, so that the price paid to merchants would in fact be 17.5 *liang*, a price much closer to the market price of 19.8 to 20 *liang* per 100 *jin*[55] (See table 5.2.)

Beginning in the late Kangxi period, because the production of Japanese copper had fallen, it was impossible for supply to meet demand. In order for trading ships from China to get their full quota, they had to wait in Nagasaki harbor. By the late Yongzheng and early Qianlong periods, there were as many as forty or more ships in this position.[56] With trading vessels unable to return on time to deliver their copper, the new acquisition of Japanese copper was consequently halted in 1738 in order to clear up accumulated debts and other problems.

According to Japanese records, in 1738 only five Chinese trading ships entered Nagasaki harbor. When trade was resumed in 1739, the number of merchant ships rose to twenty. The owners of these vessels were the so-called "self-financed merchants". Upon their return to China, Zhejiang and Jiangsu each

53 QWT, Vol. 17, Qianlong 20.

54 ZZCL, 1227–018, Qianlong 2/10/11, Zhang Qu.

55 Ibid. 1230–016, Qianlong 5/3/21; also QWT Vol. 16, Qianlong 5.

56 MQD, A72–54, Qianlong 2/5/3, Ji Zengyun; also ZZCL, 1227–009, same document.

bought one-quarter of the privately imported copper, leaving the remaining one-half for the merchant to sell freely. It is not known whether the number of owners involved was twenty. Among these twenty or so ship owners was Fan Yubin 范毓馪, who will be discussed later in this paper as an Imperial Household Department merchant.

In 1738, the Board of Revenue approved the following procedure for recruiting merchants to go abroad to acquire copper (proposed by imperial censor Zhu Fengying 朱鳳英): "If in the capital there are merchants with their own financial resources who are willing to go abroad to purchase copper, then after they provide a guarantor for the copper, they can petition the Board of Revenue, which in turn petitions the Emperor. If the merchants gain approval, officials in Zhejiang or Jiangsu are notified to allow them to go abroad for purchasing."[57]

Fan Yubin was one of those who successfully applied to the Board of Revenue for certification as a copper dealer in his status as a self-financed merchant from the capital. In 1739, the Board of Revenue approved the petition of Zu Shangzhi 祖尚志 and added: "Proposals have already been received to allow acquisition of foreign copper, and official merchant Fan Yubin is authorized to accept commissions for this task."[58] The term "official merchant" here actually signifies an official merchant of the Imperial Household Department, and not yet a formal state merchant dealing in copper. Thus, though Fan dealt in copper in the same fashion as the "quota merchants" of Jiangsu, he still had to pay customs taxes.

Let us now turn to a discussion of how the formal state merchant system was established.

In 1740, with copper coinage increasingly scarce, the provinces of Jiangsu, Zhejiang, and Fujian also petitioned to open mints, following the example of Jiangxi.[59] Because of their great distance from Yunnan, it was very time-consuming to acquire Yunnan copper, so they mostly used a combination of Yunnan and imported copper, or compulsorily purchased imported copper (that was in private market). However, as previously noted, private copper importers were entitled to keep one-half of their copper for private sale, so officials naturally could not forcibly control the entire supply. If they wanted the half remaining in private hands, they had to buy at the market price.

In 1744, therefore, "state merchants" (*guanshang* 官商) were designated to acquire copper. At that time Fan Yubin had made huge profits from copper in

57 Qianlong period, *Qinding Da Qing huidian Zeli* 欽定大清會典則例 (Imperially Endorsed, Collected Administrative Statutes and Precedents of the Great Qing) Vol. 44, 388.

58 *Qianlong chao shangyu dang*, Vol. 1, 380.

59 QWT, Vol. 16, Qianlong 5.

1740, 1741 and 1742. It also happened that he owed the government more than 1.14 million *liang* of silver. Following appeals from Naqin 訥親, in charge of the Board of Revenue, and others, the Qianlong emperor ordered that Fan meet his financial obligations through copper procurement.[60]

Each year Fan was to deliver 1.3 million *jin* of copper, broken down as follows: 300,000 *jin* to the Baoding mint in Zhili; 300,000 *jin* to the Xi'an mint in Shaanxi; 200,000 *jin* to the Suzhou mint in Jiangsu; 250,000 *jin* to the Nanchang mint in Jiangxi; and 250,000 *jin* to the Wuchang mint in Hubei. He was to deliver the copper at a price of 14 *liang* of silver (per 100 *jin* of copper) for Zhili and Shanxi, and at 13 *liang* for Jiangsu, Jiangxi, and Hubei. Copper being delivered was exempt from customs taxes. Fan was allowed to sell any surplus copper above the required quotas.[61]

Since the early Qing, most merchants that traded with Japan first received copper-acquisition funds from officials (or from Imperial Household Department merchants), and then purchased copper on their behalf. Although these copper buyers had a certain "official" character, they should not be described by the term *guanshang*, here translated as "state merchant." The system of having a specially designated state merchant handle copper acquisition really only began with the case of the Fan clan. Besides copper, the Fan clan also operated in salt and other monopoly industries.

There was always only one "state merchant." Meanwhile, beginning in 1749, the number of private merchants (also called *eshang*, or "quota merchants") was set at twelve. The reason for this is that in that year the number of Chinese trading ships permitted to go to Japan was reduced (by the Japanese authorities) to fifteen; with the Fan clan having three ships, this left twelve slots for private merchants.

Beginning in 1755, merchants could apply for licenses to show Chinese customs that the holders were authorized to import copper. From 1755 onward, except for the state merchant and "the twelve quota merchants" other merchants—including traders from Fujian and Guangdong—were not permitted to import copper from Japan.[62] (Oddly enough, the number of quota merchants was not always twelve, but the original name "the twelve merchants" stuck.) This "dual track system"—in which copper acquisition was done by the state merchant on the one hand and quota merchants on the other—was maintained until 1861.

---

60 ZZCL, 1234–007, Naqin and others; MQD, A129–35.

61 QWT, Vol. 16, Qianlong 9.

62 QWT, Vol. 17, Qianlong 20; ZZCL 1284–018 Qianlong 37/6/7, Gao Jin and others.

With regard to the identity and other details of the state merchant and quota merchants, these points have been discussed by me in a separate study and so will be omitted here.[63] Let us now move on and look at the amounts of imported copper and the procedures for its transport.

4 The Amount and Transport of Japanese Copper Imports

4.1 *The Amount of Imported Copper into China*

Scholars have long been interested in the question of how much copper was exported from Japan. It not only affected the society and economy of Qing China, but also—via Dutch traders—had social and economic implications for both Southeast Asia and Europe.

There is no complete set of records to determine Japanese copper exports to China. Appendices 1 and 2, covering the years 1663 to 1795, describe exports of copper from Japan carried by Chinese ships. Note that the figures for the years 1736 to 1754 are uncertain due to a lack of historical materials. The figures for the years 1716–1735 are extrapolated from Japanese export regulations per ship, with reference also to records and documents of trading merchants, rather than from primary materials containing the actual figures; though the extrapolations cannot be completely accurate, they should not be too far away from the actual numbers.[64] As for other figures, these are derived from the export records of major Japanese copper mines and the port of Nagasaki, though these should be seen as minimum figures (actual copper exports were probably higher).

From Japanese records as summarized in appendix 1, it can be seen that exports of copper rose sharply after the Qing lifted the ban on maritime trade in 1684. Exports were about 3–4 million *jin* per year for 1684–1695. Comparing this figure to the Qing official copper procurement quotas as shown in table 5.1, the official quota was roughly 3.8 million *jin*, meaning that Japanese exports were more or less adequate to satisfy China's minting demand. Japanese copper exports peaked in the years 1696–1710, reaching 4–7 million *jin*. At that time, the official Qing copper quota was also raised to 4–5 million *jin*.

In 1715, Japan instituted the Shōtoku Regulations and restricted copper exports to three million *jin*. This led to the situation, described earlier, of the provinces being unable to meet officially set quotas. As a result, in the

63 See Liu, "Shinnichi bōeki no yōdōshō ni tsuite: Kenryū-Kanbō ki no kanshō, minshō o chūshin ni."

64 Liu, "Kyōhō nenkan no Tōsen bōeki to Nihondō," 318.

Yongzheng period, the government lowered the amount of imported copper in the overall copper quota and partly making up for the shortfall with Yunnan copper. From all this it can be seen that, in the final analysis, the official quotas of the Qing government had to be adjusted with in response to Japanese export restrictions.

In 1742 (during the Qianlong reign) Japan further restricted copper exports, this time to 1.5 million *jin*, though actual exports hovered between 1.5 and 2 million *jin*, an average of about 1.8 million *jin* per year. The reason for these further restrictions was given in *bakufu*'s orders of 1736 and 1742 to Nagasaki to reduce the number of trading ships and the volume of trade: Domestic production was falling, causing the price of copper in Japan to rise and leading to shortfalls in copper supply. Therefore, it was deemed inappropriate for Nagasaki to sell copper relatively cheaply to China and the Dutch.[65] Thus, beginning in 1743, the formal limit on copper exports was reduced to 1.5 million *jin*.

Why did the Japanese sell copper cheaper to foreign buyers in the first place? The reason was to maximize the volume of goods traded within Japanese government-set limits on trade. Most of the Sino-Japanese trade at Nagasaki at that time was done through barter. Both sides set lower prices on their goods for trading, later selling their acquired goods at higher prices in their respective domestic markets, which allowed the Japanese side to make up their losses on copper exports. After 1721, the Japanese export price for copper was maintained at 115 *monme* (about 11.5 *liang*) of silver, well below the domestic price in Japan.[66]

Next we will look at the actual amounts of copper acquired by the state and quota merchants.

For 1736–1754, it is impossible to be very sure about the volume of Japanese copper exported to China because there are no Japanese export records. However, calculating at a rate of 100,000-plus *jin* per ship, we can estimate the amount. In 1744, when Fan Yubin was instructed to engage in copper acquisition, it was estimated that he could deliver 1.3 million *jin* per year to the various provinces. But the Japanese limit on exports was 1.5 million *jin* (a number which in actual practice varied depending upon the year of entry of copper ships into the harbor), half of which was bought up by private Chinese merchants.

---

65 *Tsūkō ichiran* [Collected Diplomatic and Trade Records], Vol. 4, 307, 354.

66 Yamawaki, Chapter "Nisshin dōbōeki no shomondai" ["Some Problems Regarding the Copper Trade Between Japan and Qing Dynasty China"] in *Kinsei Nitchū bōeki shi no kenkyū*; Ishii, "Bakumatsu ni okeru bakufu no dō yushutsu kinshi seisaku". Liu, "Kinsei Nisshin bōeki to sono kessai."

In 1744–1745, the Fan clan sent out twenty-two ships, and there were some twenty vessels belonging to private merchants. In 1746, Fan Yubin's younger brother, Fan Yuyi 范毓𩾚, brigade-commander in Zhengding, Zhili Province, petitioned to have the amount changed to 800,000 *jin*.[67] However, the trading ships were still unable to deliver the required amount of copper on time, and in 1750 the Fan quota was again lowered, to 500,000 *jin*[68] The number of ships sent out was lowered to three. Meanwhile private merchants sent twelve ships, handling 1.5–1.6 million *jin* per year.[69] Each ship carried an average 130,000–150,000 *jin* of copper.

Altogether, then, the state and private merchants were handling two million *jin* a year. This figure generally conforms to Japanese export statistics as shown in appendix 2. From this it seems clear that actual copper exports exceeded the official Japanese limits of 100,000 *jin* per ship and 1.5 million *jin* per year.

In 1765, the number of trading ships was further reduced (by Japan) to thirteen. The Fan clan as the official state merchant increased the number of ships to seven because of unpaid debts to the government. The number of private merchant ships was reduced to six. Official Qing quotas still required the state merchant to provide 500,000 *jin* (with the surplus going to meet back obligations), while private merchants supplied 800,000 *jin*.[70]

Customs documents of the time recorded the actual amounts of copper imported by the state merchant and private merchants. After the "dual-track" (state merchant plus quota merchants) system was established, trading operations with Japan gradually congregated around the single port of Zhapu 乍浦, under the jurisdiction of Zhejiang maritime customs.[71] The office of Zhapu vice prefect kept "Logs of Amounts of Copper Imported by Trading Ships for Various Years." Unfortunately, the original documents no longer exist, and we are left with only scattered information in related memorials.[72]

Because the copper acquired by the state merchants determined the amount of minting in various provinces, it was essential to control the sources of copper. Thus, beginning in 1762 it was stipulated that, in order to prevent theft and

---

67 ZZCL 1237–002, Qianlong 11/2/15, Fan Yuyi.

68 QWT, Vol. 16, Qianlong 9. The Fans delivered the copper at a slight loss, so the more they delivered, the greater their losses. This could be a major reason why they petitioned to have the quota lowered.

69 See note 62.

70 See note 62, ZZCL 1284–018, Gao Jin.

71 Liu, "Qingdai de Zhapugang yu Zhong-Ri maoyi."

72 See note 62, ZZCL 1284–018, Gao Jin. Also, Matsuura Akira, "Shindai zenki no Sekkō Kaikan to kaijō bōeki," introducing the existing "Yellow Books" (customs accounts) from Zhejiang maritime customs for the years Yongzheng 1, Yongzheng 5, and Qianlong 2.

illegal sale, after the copper was brought back to China, the governors of the provinces through which copper was transported would have to file reports in time according to the "Regulations for Transporting Copper and Lead to the Capital" of 1749.[73] Consequently, we can establish the amount of copper imported by the state merchant by referring to reports filed by the governor of Zhejiang (which are available for 1762–1793, except for 1781 and 1784).

As for the private merchants, besides the fact that Jiangsu and Zhejiang purchased 50 percent of their shipments, beginning in 1744 Jiangxi also purchased 10 percent.[74] That is, 60 percent of their copper was bought by provinces, while the merchants could privately sell the remaining 40 percent. Because this was private copper, besides customs registries, the various provinces did not report their relevant records to the capital. Currently, only records on the volume of imported copper from 1767 to 1771 exist.[75]

Please refer now to tables 5.3 and 5.4.

The amount of copper imported by the state merchant can be seen in table 5.3. The reports of the governor of Zhejiang uniformly declare amounts of 100,000 *jin* (1,000 cases) per ship. Except for the year 1765, when one ship carried a double load (having picked up the copper of another ship that had run aground),[76] the amount did not change at all right up until 1783, when state merchant Fan Qingji 范清濟 went bankrupt and Wang Shirong 王世榮 was given the job. Changes only began after Wang, and later Qian Mingcui 錢鳴萃, took responsibility as the state merchant. Unfortunately, the reports filed after 1785 are in much less detail, and indicate only the total amount of imported copper, but omit the number of ships and the ship owners' names.

Although in principle an export volume of 100,000 *jin* per ship was stipulated, in fact, as noted earlier, each ship carried more than that. Comparing private ("quota") merchants with the state merchant, the amount of cargo per quota-merchant ship between 1767 and 1771 surpassed that per state-merchant ship; each of the former carried an average of 146,000 *jin*. As table 5.5 shows, private merchant ships blown off course to Fujian also mostly carried more than 100,000 *jin* of copper.

As a result, it is possible that state merchants' reports of their copper imports were merely pro forma, and one might suspect the figures to be incomplete, with the surplus of unreported copper being sold for a profit in the open market. However, it is also possible that because the Fan clan had large debts, it

73 *Qianlong chao shangyu dang*, op. cit. (note 47) Vol. 3: 812–13.

74 QWT Vol. 16, Qianlong 9.

75 See note 62, ZZCL 1284–018, Gao Jin.

76 See Matsuura, "Shindai kanshō saiben yōdō benkai senseki," 52.

TABLE 5.3 *Imports of Japanese copper by state and quota merchants* (*unit:* jin)

Year	Identity of merchants	Number of ships	Copper (in *jin*)	Sources	
1762	Fan Qingzhu	3	300,000	GDQZ	16–656
1763	"	3	300,000	ibid.	20–260
1764	Fan Qinghong	4	400,000	ibid.	23–734
1765	Fan Qingji	4 (5)	472,100	ZZCL	1265–006
1766	"	9	900,000	ibid.	1268–016
1767	Fan Qingji	5	500,000	GDQZ	29–255
	*Quota merchant ships	6	839,900	ZZCL	1284–018
1768	Fan Qingji	5	500,000	ZZCL	1270–004
	*Quota merchant ships	6	940,000		1284–018
1769	Fan Qingji	5	500,000	ibid.	1273–028
	*Quota merchant ships	6	870,000		1284–018
1770	Fan Qingji	6	600,000	ibid.	1278–025
	*Quota merchant ships	7	1,107,700		1284–018
1771	Fan Qingji	6	600,000	ibid.	1282–012
	*Quota merchant ships	5	623,500		1284–018
1772	Fan Qingji	5	500,000	ibid.	1287–002
1773	"	6	600,000	GDQZ	33–555
1774	"	7	700,000	ZZCL	1292–002
1775	"	8	800,000	ibid.	1298–015
1776	"	6	600,000	ibid.	1303–029
1777	"	6	600,000	GDQZ	41–151
1778	"	7	700,000	ibid.	45–801
1779	"	7	700,000	ZZCL	1308–008
1780	"	7	700,000	ibid.	1314–004
1781	"				
1782	"	6	600,000	GDQZ	54–169
1783	Fan Qingji	1	100,000	ibid.	58–554
	Wang Shirong	1	100,000	ibid.	
1784	Wang Shirong	?	1,350,000	ZZCL	1318–029
1785	"	?	560,000	ibid.	1322–033

Year	Identity of merchants	Number of ships	Copper (in *jin*)	Sources	
1786	"	?	1,320,300	GDQZ	62–739
1787	"	?	420,000	ibid.	66–743
1788	Qian Mingcui	?	818,900	ibid.	70–696
1789	"	?	1,131,700	ibid.	74–485
1790	"	?	640,000	ZZCL	1330–008
1791	"	?	700,000	ibid.	1334–041
1792	"	?	700,000	ibid.	1338–005
1793	"	?	700,000	ibid.	1341–023

TABLE 5.4 *Number of ships dispatched and amount of copper purchased for state merchant and quota merchants, 1767–1771*

Year	Chinese records			Japanese records	
	Ships		Copper (*jin*)	Ships	Copper (*jin*)
1767	State merchant	5	500,000		
	Quota merchants	6	839,900		
	Subtotal	11	1,339,900	12	1,512,635
1768	State merchant	5	500,000		
	Quota merchants	6	940,000		
	Subtotal	11	1,440,000	11	1,523,746
1769	State merchant	5	500,000		
	Quota merchants	6	870,000		
	Subtotal	11	1,370,000	11	1,280,565
1770	State merchant	6	600,000		
	Quota merchants	7	1,107,700		
	Subtotal	13	1,707,700	15	1,865,588
1771	State merchant	5	600,000		
	Quota merchants	6	623,500		
	Subtotal	11	1,223,500	14	1,516,197
Total	State merchant	27	2,700,000		
	Quota merchants	30	4,381,100		
	Total	57	Total 7,081,100	63	7,699,131

TABLE 5.4 *Number of ships dispatched and amount of copper* (cont.)

Year	Chinese records			Japanese records	
	Ships		Copper (*jin*)	Ships	Copper (*jin*)
Average per year		11.4	1,416,220	12.6	1,539,826
Average per ship	State merchant		100,000		
	Quota merchants		146,037		
	Average		124,230		122,208

was unable to come up with the capital to do more than acquire the stipulated amounts.[77] Also, because the compulsory purchase price for copper acquired by the state merchant was only 13.5 *liang* of silver (including transport costs to the various provinces), there was no profit to be made in bringing back an excess. It was perhaps preferable to bring back other commodities and sell these, to make up for the losses suffered on the copper.

Private merchants, on the other hand, could sell 60 percent of their copper to the three provinces of Jiangsu, Zhejiang, and Jiangxi at an official price of 17.5 *liang*, without being responsible for transport to distant regions, and the remaining 40 percent of the copper, which they retained, could be sold at the even higher open market price. Thus private merchants did not lose out by bringing in more than 100,000 *jin* per ship.

In 1763, according to Japanese records, ship owner Wang Lüjie 王履階 of the Twelve Families 十二家 (another name of the quota merchants) presented the Japanese side with a new proposal. Each year China would export 300 *kan* 貫 (30,000 *liang*) of silver in bulk. In return Japan would ship 300,000 *jin* of copper (at the special price of 10 *liang* per 100 *jin*). Thereafter China would export various metals including gold and foreign silver (usually Mexican silver dollars) for the *bakufu* to use for minting currency. Besides copper, the Japanese side would also increase exports of maritime products and other goods. This became known as *betsudan shōhō* 別段商法 (temporary trade exceptional

77 See Wei and Wu, "Qingdai zhuming huangshang Fan shi de xingshuai," 140–43. Also, Liu, "Shinnichi bōeki no yōdōshō ni tsuite: Kenryū-Kanbō ki no kanshō, minshō wo chūshin ni," 128–29.

quota outside the regular trade, in order to obtain special commodities).[78] Variously named commercial arrangements and contracts were used up until the middle of the 19th century. Naturally, the partners in these arrangements included the state merchants. Therefore, on the surface it appears as if the amount of trade was constantly shrinking; in fact, though the number of ships and volume of copper traded steadily diminished, the total amount of bilateral trade did not decline.

In 1784, state merchant Wang Shirong increased the number of ships dispatched, in order to compensate for the shortfall in copper deliveries by Fan Qingji as well as to deliver his own quota of 1.45 million *jin*. At this point, as shown in appendix 2, Japanese records indicate a sudden increase in exports to more than 2.3 million *jin*. As noted earlier, by comparing Qing documents with Japanese export records, it can be seen that the amount of copper imported from Japan during the Qianlong era essentially matches, or even exceeds, Japanese records of copper exports. (Japanese records may not include *betsudan shōhō* exports or smuggled copper.) Right up until the late Qianlong period, annual imports of Japanese copper stayed at about 1.4 to 1.5 million *jin*.

4.2 *Transport of Imported Copper*

Before the Qianlong period, as noted earlier, the shortage of copper prohibited provincial mints from operating regularly. Thus, copper transport prior to the Qianlong era refers mainly to supplying the capital. After purchasing copper brought back from Japan to the Zhejiang-Jiangsu coast, provincial copper-procurement officials or Imperial Household Department merchants would ship it to the capital via the Grand Canal.

After the copper administration reforms of the early Qianlong era, in order to ameliorate the problem of the rising cost of copper coinage in various provinces, provincial mints were opened one after another, increasing the demand for copper. Some provinces exclusively used Yunnan copper, others exclusively imported copper, but most relied on some combination of the two. Provinces that used imported copper in the Qianlong reign included Zhili, Shanxi, Jiangsu, Zhejiang, Shaanxi, Hubei, Jiangxi, Fujian, and Guangdong (nine in all). Using which copper source depended upon the province's location, though, besides problems of ease or difficulty of acquisition and transport, the market value of the copper coinage and the costs of minting also had to be considered.

Fujian and Guangdong only used imported copper in the early Qianlong era. Guangdong originally purchased a small amount of copper from Western trading vessels; but because the supply was unstable, after 1756 Guangdong

78 Nakamura, *Kinsei Nagasaki bōeki shi no kenkyū*, 423–59.

exclusively used Yunnan copper. Fujian was occasionally able to purchase Japanese copper from ships that had been blown off course on their return to China and ended up in Fujian (see table 5.5). But after 1770, by order of the imperial court, not only were Fujianese merchants banned from going to Japan to acquire copper, the province was even prohibited from buying copper from ships accidentally blown off course to Fujian. Thus the province had no choice but to purchase copper for minting from faraway Yunnan.[79]

TABLE 5.5 *Copper trading ships blown off course to Fujian, and copper purchases by Fujian, in the Qianlong Reign*

Year	Owner	Copper cargo (*jin*)	Fujian purchases (*jin*)	Sources
1742	Xu Weihuai	77,499 (recovered)	* 70,420	*Lidai Baoan* 4–2377
1744	Lin Junji	244,100	* 118,933	ZZCL 1234–005
1745	You Zhongmou	138,478 *	138,478	ibid. 1236–022
1746	Chen Tianyuan	109,000	* 104,957	ibid. 1237–015
1748	Chen Xin	99,000	* 96,120	ibid. 1239–006
1749	Wu Sizhe	more than 119,000	* 107,402	ibid. 1240–007
				1241–009
1751	Chen Mingchuan	unknown	unknown	GDQZ 2–688
1751	Gao Rixin	unknown	unknown	ibid.
1752	Qian Mingcui	208,640	* 121,538	ibid.
1753	Wu Xiuruo	more than 151,000	* 148,242	ibid. 5–861
1756	Gao Shanhui	172,500	172,500	ibid. 16–238
1757	Xia Lüduan	126,585 *	126,585	ZZCL 1249–013
1763	Wang Zaoyang	more than 168,000 *	* 120,000	GDQZ 20–293
1770	Zheng Kongyang	85,320 *	85,320	ZZCL 1276–008

* Amount calculated at 97 percent purity

79 ZZCL 1290–011, Qianlong 38/8/19, Zhong Yin. Also Matsuura, "Nagasaki bōeki ni okeru Kōsetsushō to Binshō."

The demand for foreign copper among the provincial mints changed depending upon the overall amounts imported, and was not fixed. In the *Qinding Da Qing huidian* 欽定大清會典 (*Collected Statutes of the Great Qing Dynasty*), Volume 14, printed in 1818, the total amount of imported copper used by the various mints was 1.26 million *jin* (see table 5.6). There should not be a great difference between this figure and the figure for the late Qianlong reign, and we may take it as a maximum level of demand for minting. This total accounted for about 12 percent of total copper supply in China. Some of this copper was sourced from the state merchant, but most of it came from private merchants.

After the copper quota for the state merchant (then the Fan clan) was lowered to 500,000 *jin* in 1750, there is no indication in historical records as to how much of this was sent to the various provinces. Based on reports of the shipment of state-merchant copper for various years, we can know the accurate figures (excluding compensatory copper and extra copper), as follows:

Zhili Baozhi Mint	250,000 *jin*
Jiangsu Baosu Mint	55,906 *jin*
Zhejiang Baozhe Mint	50,000 *jin*
Jiangxi Baochang Mint	50,000 *jin*
Hubei Baowu Mint	50,000 *jin*
Shaanxi Baoshaan Mint	50,000 *jin*

The total for the six provinces is 505,906 *jin*. There was no change in this figure right through to the Jiaqing reign (1796–1820).[80] When the state merchant brought copper back to China, he had the responsibility for delivering it to the various mints. Private merchants, on the other hand, sold directly from their ports of return in Zhejiang and Jiangsu to the provinces of Zhejiang, Jiangsu, and Jiangxi. Shanxi Province, which used exclusively imported copper, had to recruit merchants to go to Suzhou or Hankou to buy copper and transport it back. The area in which imported copper circulated was not limited to the coastal provinces, but also included Shanxi and Shaanxi, in the far interior of the country.

Yunnan copper was mainly transported via the Yangzi River waterway and the Grand Canal; the situation was similar for imported copper. Based on distance and routing of the shipments, we can distinguish between the Zhili route

80 Guangxu period, *Qinding Da Qing huidian shili* 欽定大清會典事例 [Imperially Endorsed, Collected Administrative Statutes and Precedents of the Great Qing] Vol. 218, "Hubu Qianfa" ["Board of Revenue Monetary Law"], Jiaqing 2.

TABLE 5.6 *Types and amounts of copper used by various mints in the Jiaqing period*

Mint name	Type of copper (unit = *jin*)			
	Yunnan copper	Japanese copper	Hankou copper	Local copper
Board of Revenue Baoquanju	4,106,880			
Board of Works Baoyuanju	1,945,406			
Zhili Baozhiju		290,000		
Shanxi Baoqinju		120,000		
Jiangsu Baosuju	216,666	250,000		
Zhejiang Baozheju	400,000	400,000		
Jiangxi Baochangju	170,000	120,000		
Hubei Baowuju	234,000	40,000	100,000	
Shaanxi Baoshaanju	350,000	40,000		
Sichuan Baochuanju				600,000
Hunan Baonanju	200,000			100,000
Fujian Baofuju	330,000			
Guangdong Baoguangju	176,000			
Guangxi Baoguiju	106,250			
Yunnan Baoyunju	555,000			
Dongchuanju	183,950			
Guizhou Baolingju	341,700			
Dadingju	93,900			
Total	9,409,752	1,260,000	100,000	700,000

Source: Jiaqing Da Qing huidian [Imperially Endorsed, Collected Administrative Statutes of the Great Qing] (1818) Vol. 14, Board of Revenue, "Monetary Laws"

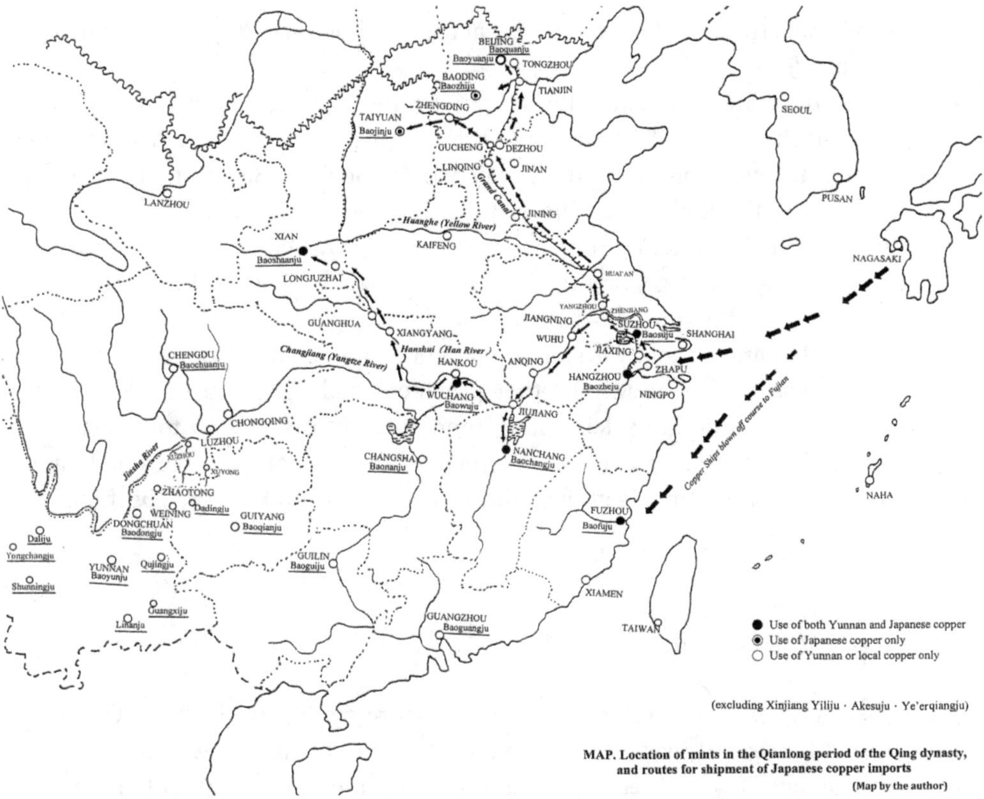

MAP 5.1 *Location of mints in the Qianlong period of the Qing Dynasty and routes for shipment of Japanese copper imports*

(the Grand Canal) and the Shaanxi-Hubei-Jiangxi route (the Yangzi); the route to Shanxi was probably largely the same as for Zhili.[81] (see Map 5.1)

In the 1750s, copper importing was concentrated at the port of Zhapu, in Pinghu County, Jiaxing Prefecture, Zhejiang Province. One reason for choosing Zhapu as the main base for trading with Japan was its proximity to Japan, making for easier sailing, and also its proximity to the Grand Canal, along which imported copper was shipped to the Baosu Mint in Suzhou, from where it was shipped to other destinations. But perhaps an even more important reason is that by concentrating the copper trade with Japan in a single port, it was easier for the authorities to supervise the merchants and control the copper. Also,

81 See Liu, "Qing Kangxi-Qianlong nianjian yangtong de jinkou yu liutong wenti," 118–26.

Zhapu was the site of a Manchu naval base and played a key role in maritime defense.[82]

For the state merchant, after his ships entered port, the Zhejiang authorities had the responsibility to verify the amount of copper imported and also to monitor its shipment to the Baosu Mint in Suzhou. Beginning in 1762, the authorities also had to file reports indicating the dates of entry into the province, of shipment, and of departure from the province; this was intended to reduce delays, sinkage, and theft. At the same time Zhejiang Province had to notify Jiangsu Province in writing.

For Zhejiang's 50,000 *jin*, state-merchant copper could be delivered directly to the nearby Baozhe Mint. Other provinces had their copper first sent to Suzhou for temporary safekeeping, after which it was sent on to its ultimate destination. After 1797, the four provinces of Zhili, Shaanxi, Hubei, and Jiangxi switched to a system of sending officials to Suzhou to take charge of shipment from there.[83]

5 Conclusion

There was extensive reform of copper administration in the early Qianlong period for several reasons: Against a background of shortages of official copper coinage, it was necessary to acquire a large amount of copper to mint. But the merchants recruited by the government for this task could not fulfil their quotas, leading to a shortage of copper at the metropolitan mints. In addition, merchants accumulated huge shortfalls and debts to the government, and as the situation became more severe, something had to be done to put the system on a sound footing.

In 1738, acquisition of copper from Japan was temporarily halted, in order to clear accumulated debts and related problems. During this halt, Yunnan copper was used to replace imported copper for minting purposes. It was decided that Yunnan on the one hand, and the two provinces of Jiangsu and Zhejiang on the other (the latter two having formerly imported Japanese copper), would each be responsible for delivering half of the capital's needs. However, given the great distance between Yunnan and the latter two provinces and the urgent need in the capital for copper, it was decided that all the copper for the capital would be shipped directly from Yunnan. Jiangsu and Zhejiang, meanwhile, went back to buying Japanese copper.

82 Liu, "Qingdai de Zhapugang yu Zhong-Ri maoyi," 225.

83 Guangxu period, *Qinding Da Qing huidian shili*. Vol. 218, "Hubu Qianfa," Jiaqing 2.

At that time, the production of copper in Yunnan had already been rising as a result of management changes in the late Yongzheng reign. After the Qianlong emperor ascended the throne, he ordered that production be increased even faster. With the supply of imported copper falling, beginning in 1739, all of the capital's copper requirements were to be fulfilled from Yunnan in order to stabilize currency minting. Thus the capital's shift from Japanese copper to Yunnan copper was not pre-planned, but was made inevitable by the decline in copper imports and the rise in production of Yunnan copper.

After 1739, in order to clear up accumulated shortfalls, the acquisition of imported copper was completely turned over to private merchants, who traded using their own capital. The state purchased one-half of the copper these merchants brought back, and the merchants were allowed to sell the other half on the open market. Imported copper thus circulated widely. This new system not only created incentives for merchants to buy copper for the state, it also helped stabilize copper prices.

Later, as the copper-currency-shortage problem worsened, provinces were allowed to open their own mints, but then the supply of copper again became a problem. As a result, it was decided that Imperial Household Department merchant Fan Yubin—who owed the government huge sums of money—would act as a state merchant acquiring copper, in order to meet his obligations. The state merchant system for acquiring copper was not, as is generally believed, enormously profitable, but rather was one in which profits from salt were used to compensate for losses incurred from dealing in copper.[84]

The establishment of the "dual-track" system of state merchant and "quota merchants" shows that the Qing tightened control over merchants, in order to control the supply of copper. After the middle of the Qianlong period, however, because of natural disasters and the falling value of copper coinage, the use of salt revenues to make up copper losses ceased to be effective, and the Fan clan went bankrupt.

During the Qianlong period, although the amount of copper imported from Japan declined, so that it only accounted for about 12 percent of the total national supply, it retained certain advantages in terms of purity and convenience of delivery, so that it still played a pivotal social and economic role in the Qing Dynasty. In 1769, Grand Secretary Chen Hongmou 陳宏謀 petitioned for a halt to the purchase of imported copper. After review by several top ministers, the petition was rejected, and it was concluded:

84 *Qing shilu* [Veritable Records of the Great Qing] Vol. 1172, Qianlong 48/1/renyin (1783/2/11).

> "We have found that trading merchants handle more than 960,000 jin of copper for us each year. Along with Yunnan copper, this is distributed to the capital and various provinces for minting. If the importing of copper is halted completely, there will not be enough for distribution."[85]

There are many other interesting problems associated with copper that have not been touched on here for reasons of space and a lack of historical materials. These include such issues as the relationship between minting and copper supply and demand in various locations, and the relationship between private copper circulating among the public and the prices of copper and other commodities. These are too broad to address here and will have be left for future consideration.

APPENDIX I *Copper exported from Japan by Chinese ships (1663–1715)*

Year	Amount exported (*jin*)		Number of ships
	A Japanese records	B Dutch records	
1663	453,700		29
1664	249,860		38
1665	197,300		36
1666	451,404		37
1667	784,840		33
1668	830,200		43
1669	492,200		38
1670	732,470		36
1671	1,351,130		38
1672	1,158,100		43
1673	1,096,650		20
1674	1,127,090 (1,032,600)	1,831,900	22
1675	1,921,640 (1,935,400)		29
1676	1,513,472	1,044,200	24
1677	1,558,234	1,200,000	29

85 Ibid. Vol. 849, Qianlong 34/12/renshen (1770/1/20).

Year	Amount exported (*jin*)		Number of ships
	A Japanese records	B Dutch records	
1678	1,641,505	1,800,000	26
1679	1,993,100	1,847,770	33
1680	1,593,600		29
1681	536.700	* 327,875	9
1682	3,021,850	3,283,925	26
1683	2,329,355	* 2,825,356	27
1684	2,614,888 (2,675,100)	1,946,300	24
1685	2,328,440 (3,288,200)	2,091,400	73
1686	3,244,493.5 (4,455,700)		83
1687	4,294,562 (3,830,200)	3,385,300	115
1688	3,921,730 (3,370,600)	2,766,000	117
1689	3,542,622.5 (3,352,568.5)	5,735,200	70
1690	3,743,873 (3,766,873.5)		70
1691	3,220,955 (2,939,440)	3,400,000	70
1692	3,546,374 (2,564,650)	3,770,400	70
1693	3,362,685 (3,502,717)	3,744,800	70
1694	3,440,799 (3,298,150)		70
1695	4,161,136 (4,375,762)	7,406,800	60
1696	7,019,768 (6,946,589)		70
1697	6,408,178 (6,325,035)		70
1698	6,082,395	6,402,000	67

APPENDIX I *Copper exported from Japan* (cont.)

Year	Amount exported (*jin*)		Number of ships
	A Japanese records	B Dutch records	
1699	5,236,435	2,026,400	69
1700	3,629,515	1,808,400	53
1701	3,833,051		56
1702	3,811,372		80
1703	3,943,032		80
1704	5,343,315		80
1705	4,967,641		80
1706	5,100,356		80
1707	5,101,460		80
1708	6,603,994		59
1709	5,170,521.5		54
1710	4,955,261.5		51
1711	4,257,850		57
1712	2,938,599 (2,778,013)		59
1713	3,909,199 (3,723,187)		40
1714	3,999,915 (3,853,100)		51
1715	763,731 (755,581)		7

Source:
A Sumitomo shūshishitsu, *Sen'oku sōkō* Vol. 9, Furoku "Dōyushutsutakahyō."
B Iwao, "Kinsei Nisshi bōeki ni kansuru sūryō teki kōsatsu," 1015.
* Nagazumi, *Tōsen yushutsunyūhin sūryō ichiran*, 1637–1833.
** *Nagasaki jitsuroku taisei* and *Tsūkō ichiran*; note that Japanese and Dutch statistics differ (see note 65).

APPENDIX II *Copper exported from Japan by Chinese ships (1716–1796)*

Year	Amount exported (*jin*)	Number of ships	Notes
1716	720,000	7	
1717	4,370,000	43	
1718	4,100,000	40	
1719	3,670,000	37	
1720	2,956,000	36	
1721	2,696,000	33	
1722	2,628,000	33	
1723	2,736,000	34	
1724	976,000	13	
1725	2,528,000	30	
1726	3,540,000	41	
1727	3,480,000	42	
1728	1,627,000	22	
1729	3,040,000	31	
1730	3,528,000	38	
1731	3,590,000	38	
1732	3,440,000	36	
1733	2,656,000	28	
1734	2,794,000	31	
1735	2,806,000	29	
1736		16	
1737			5
1738			5
1739			20
1740			25
1741			14
1742			12
1743			15
1744			20
1745			20
1746			10
1747			10
1748			12
1749			13
1750			10

APPENDIX II *Copper exported from Japan* (*cont.*)

Year	Amount exported (*jin*)	Number of ships	Notes
1751			11
1752			15
1753			25
1754			24
1755	1,869,729	16	12
1756	1,862,738	14	7
1757	1,714,985	11	12
1758	2,148,013	14	14
1759	2,097,798	15	20
1760	1,847,638	11	12
1761	2,453,885	16	12
1762	1,733,698	13	15
1763	1,824,163	12	13
1764	1,712,505	14	14
1765	1,648,262	13	12
1766	1,901,893	16	12
1767	1,512,635	12	13
1768	1,523,746	11	9
1769	1,280,565	11	13
1770	1,865,988	15	13
1771	1,516,197	14	13
1772	1,408,505	11	13
1773	1,352,483	11	13
1774	1,933,596	16	13
1775	1,600,977	12	13
1776	1,581,172	12	13
1777	1,254,276	12	13
1778	1,447,577	12	13
1779	1,528,218	13	13
1780	1,354,218	12	13
1781	1,606,843	13	13
1782	1,535,981	13	13
1783	1,336,283	10	13
1784	2,338,334	16	14
1785	1,811,643	13	15

Year	Amount exported (*jin*)	Number of ships	Notes
1786	1,645,451	13	12
1787	1,384,899	12	13
1788	1,879,647	15	13
1789	1,277,858	11	13
1790	1,103,437	11	9
1791	1,304,062	13	10
1792	1,203,750	12	10
1793	1,219,800	13	(Ships outside officially authorized number: 2)
1794	1,103,437	11	10
1795	902,812	9	10

Sources:
(1) 1716–1735: Liu, "Kyōho nenkan no Tōsen bōeki to Nihondō" (see note 64).
(2) 1755: *Suijinroku* in *Katsu Kaishū zenshū*, Vol. 9, 26–49.
(3) For the number of ships, see Mitsui and Doi, *Shin Nagasaki Nenpyō*. The first number of ships refers to ships entering Nagasaki, the second to those leaving the port.

CHAPTER 6

The Trade in Dried Marine Products from Nagasaki to China during the Edo Period

Matsuura Akira 松浦章

1 Introduction

The Chinese intellectual Wang Tao 王韜 (1828–1897), who visited Japan at the end of the Qing Dynasty, described the restricted nature of contact between Japan and the Qing Empire in the fourth volume of *Wengyou yutan* 甕牖餘談, under the title "Tongshang Riben shuo 通商日本説 [Tales of Trade with Japan]". The relationship between China and Japan during most of the Qing Dynasty, he observed, was limited to trade conducted by Chinese junks at Nagasaki 長崎.[1] The most important goal of Chinese traders bound for Nagasaki was to acquire copper for Chinese coinage. In exchange for this copper, Chinese merchants transported Chinese silk products and raw silk, as well as sugar and Chinese medicines. In order to legally trade at Nagasaki,

* Translator's note: I use "dried goods" for *tawaramono* 俵物, goods that were traditionally transported and stored in straw bags. The precise goods varied, but in this context generally were understood to be dried marine products, and above all the three discussed here.

** Paper translated by Michael Thornton.

1 日本一國、密邇中土、・・・國朝順治以後、惟通市、不遣使、其市亦惟中國商船往、無倭船來也。其與中國貿易在長崎島、百貨所聚、商旅通焉。國饒銅、中土鼓鑄所資、自滇銅而外、兼市洋銅、・・・毎年採購定額、四百四十三萬餘斤、設官商額船十六艘、皆以內地綢緞・絲棉・糖・薬往易、商辦銅斤、必藉倭照以爲憑驗。The country of Japan is very close to China.... Once the country's government established control, only trade occurred, with no diplomatic envoys. Trade only took place with Chinese merchant ships, with no Japanese ships coming [to China]. Their trade with China took place on the island [sic] of Nagasaki where all manner of goods were gathered and traded by companies of travelling merchants. The country was abundant in copper, and supplied the coin minting needs of China: in addition to copper from Dian (Yunnan 雲南), foreign [Japanese] copper) was purchased.... Every year a set amount was acquired—some 4,430,000 *jin* 斤. The position of state merchant was established, and sixteen vessels carrying domestically produced silk brocade, cotton, sugar and medicines travelled to trade, acquiring copper ingots. This trade depended upon the required Japanese Trade Licence. *Biji xiaoshuo daguan*, 258.

 | DOI 10.1163/9789004304512_007

merchants required a seal issued by the Japanese authorities, the so-called 'Japanese seal' or 'Japanese Trade License (*shinpai* 信牌)'.

Wang's account lists several characteristic products traded between China and Japan during the Edo period (1603–1867). The largest Japanese export to China was Japanese-mined copper. Additionally, in 1875 He Ruzhang 何如璋 (1838–1891), then Chinese envoy to Japan, noted in his *Shidong zaji* 使東雜記 [Miscellany of an Envoy to Japan] that "many Chinese merchants take raw cotton and white sugar, and return with various marine products such as sea cucumber (*iriko* 煎海鼠) and dried abalone (*hoshiawabi* 干鮑)."[2] His description indicates that Japan's primary exports to China at the beginning of the Meiji period (1868–1912) were also copper and various dried marine products.

The 1828 woodblock book *Zhapu beizhi* 乍浦備志 [Supplementary Gazetteer of Zhapu] recounts the history of the market town of Zhapu in Pinghu district of Jiaxing prefecture in Zhejiang (浙江省嘉興府平湖縣乍浦鎮). Volume 14, which records events of the mid-Qianlong years (1736–1795) when Zhapu was the base of trade with Japan, describes how merchants were organized: The Chinese side comprised two groups of merchants, one of state officials and one of private citizens. Each had three ships, for a total of six sent to Japan every year between the summer solstice 夏至 and the minor heat 小暑 of the old calendar (corresponding roughly to late June and early July in the Gregorian calendar). Those ships carried sugar from Fujian 福建 and Guangdong 廣東, as well as Chinese goods highly desired by Japanese, and set out eastwards for Japan.[3] Then, in the ninth month the Chinese ships were loaded with copper, kelp, dried sea cucumber and various marine plants before returning to Zhapu, revealing that, in addition to copper, marine products such as kelp and sea cucumber formed an important component of the trade with Japan.[4]

During the Edo period, the quantity of dried sea cucumber, dried abalone and shark fin (*fukahire* 鱶鰭), exported from Nagasaki to China grew considerably. Particularly during the Qing Dynasty, the growing popularity of seafood produced a taste for sea cucumber (*haishen* 海參), dried abalone (*fuyu* 鰒魚) and shark fin (*yuchi* 魚翅) in China. As a result, export of these marine products increased steadily during the Genroku period 元禄時代 (1688–1703; Kangxi 康熙 27–42). Consumption was greatest in the Lower Yangzi delta region, but by the late Qing Dynasty consumption increased as dried sea

2 Wang, *Xiaofanghuzhai yudi congchao*, 8001.

3 *Zhongguo difangzhi jicheng xiangzhenzhi zhuanji*, 229.

4 In the middle of the ninth month, the ships were loaded with copper ingots, kelp, sea cucumber and other marine plants before returning to Zhapu. Ibid., 230.

cucumber, dried abalone and shark fin made their way into the interior, enlivening dining tables around China.[5]

Qing China's most sought-after product from Japan was Japanese copper, known in China as 'foreign copper'.[6] The amount of copper supplied to the Qing government in 1716 totaled 4,130,000 *jin* 斤, of which 'foreign copper' comprised some 2,770,000 *jin* and domestically produced Yunnan copper some 1,660,000 *jin*.[7] Proportionally, Japanese copper represented 62.5 percent of Qing China's supply, compared to 37.5 percent from Yunnan 雲南, demonstrating that Chinese domestic production could by no means satisfy the needs of Qing society. Therefore the import trade from Nagasaki, which was also relatively convenient, held considerable value for Qing society.

The vessels operating that trade were junks, Chinese merchant sailing ships that Edo period Japanese referred to as *Tōsen* 唐船. In order for these ships to sail safely and smoothly across the open ocean, they required large quantities of cheap and portable material for ballast. Sugar fulfilled the criteria of cost and quantity, and moreover proved popular in Japanese society.[8] In return went the 'foreign copper,' the Japanese copper sought by the Qing court. However, as Japanese copper production gradually declined, export of dried marine products increased to supplement the metal trade.

In this essay, I wish to investigate historical changes in the Edo period trade in copper and dried marine products at Nagasaki using the logs and cargo inventories of ships returning from Nagasaki to China.

2 Trade in Dried Marine Products with China during the Genroku, Hōei 宝永 and Shōtoku 正德 Periods

After pacifying the Ming loyalist Zheng 鄭 clan of Taiwan, the Qing proclaimed the 'Order of the Open Seas' (*zhanhailing* 展海令) in 1684, prompting large numbers of Chinese coastal merchants to set sail overseas. They focused particularly on Japan. As a result, Nagasaki hosted several thousand Chinese

5 Matsuura, "Nisshin bōeki ni yoru tawaramono no Chūgoku ryūnyū ni tsuite", 19–38; Matsuura, *Shindai kaigai bōekishi no kenkyū*, 382–402.

6 Yan, *Qingdai Yunnan tongzheng kao*, 3. The treasury of the Board of Finances and the mint of the Board of Works in Beijing required over 4,400,000 *jin* of copper every year to mint coins. Ibid., 4.

7 *Shangyu Tiaoli*, hubu, "bantong tiaoli" 上諭條例 戸部、辦銅條例 [*Edicts and Regulations*, Finances Section, "Regulations for the procurement of copper"], 1736; Matsuura, *Edo jidai Tōsen ni yoru nitchū bunka kōryū*, 111.

8 Matsuura, "Edo jidai Tōsen ni yoru satō yu'nyū to kokunai shōhi no tenkai," 335–59.

residents year round, leading to restrictions on Chinese settlement in the city proper and the development of so-called Chinese quarters (*Tōjin yashiki* 唐人屋敷, known by the Chinese as *Tang-guan* 唐館) where the Chinese were forced to reside.[9] This period also saw the start of a steady increase in exports of dried marine products to China.[10]

On that note, let us look at the *Tō tsūji kaisho nichiroku* 唐通事會所日録 [Daily Record of the House of Chinese Interpreters] for a sense of the state of dried marine product exports in the Genroku, Hōei and Shōtoku periods (1688–1716), when some seventy to eighty trade junks arrived in Nagasaki from China every year.

2.1 *Dried Abalone and Shark Fin*

Sea cucumber and abalone figured prominently in the request for goods by the Chinese residing in Nagasaki's Chinese quarters on 1689/4/3. Merchants planned to dry sea cucumbers and abalone every day in order to return to China with them.[11]

A passage dated 1698/9/2 records that the twenty-eighth ship[12] returning to China carried a large amount of copper, which had made the ship too heavy to sail, forcing it to return to Nagasaki.[13] The ship carried "about 1,500 crates of copper, along with dried goods, together forming a large load."[14] On 1702/6/8, the twenty-third Ningbo 寧波 ship of 1701 was sent to Nagasaki after drifting ashore on Amakusa 天草. The Ningbo ship was loaded with dried marine products, but bad weather delaying its departure from Nagasaki had caused the

---

9 Matsuura, *Kinsei Higashi Ajia kaiiki no bunka kōshō*, 31–50.

10 Matsuura, *Shindai kaigai bōeki shi no kenkyū*, 382–402.

11 いりこ・あわひ類毎日干候て連〻買渡ル儀ニ御座候、蔵に入置候而ハ、尤時々干可申候。 [Sea cucumbers and abalone are dried daily, and sold seemingly without end. They are in warehouses, and it must be said that an impressive amount is always dried.] *Tō tsūji kaisho nichiroku*, 1: 244.

12 Chinese ships entering Nagasaki were assigned numbers according to the order of their arrival. In principle, this numbering system (known as *banzuke* 番付) included the zodiac sign of the lunar calendar year; for example, ships arriving in the year of the dragon would receive appellations such as "Year of the Dragon, first ship, Year of the Dragon, second ship," and so forth. Extraordinary ships, such as those shipwrecked on Japanese shores, were not numbered. Instead, they were listed as extras (*bangai* 番外), thus "Year of the Dragon, Extra." [Translator's note: I have omitted the zodiacal year in the ship's numbers below, relying instead on the Western year approximation.]

13 *Tō tsūji kaisho nichiroku*, 2: 330.

14 Ibid., 2: 331.

FIGURE 6.1 *Dried abalone.*
COURTESY OF NAGASAKI MUSEUM OF HISTORY AND CULTURE (長崎歴史文化博物館)

sea cucumber and abalone to go damp.[15] In 1706/7, the twenty-second Nanjing 南京 ship of 1705, which had become stranded in Satsuma 薩摩, was towed to Nagasaki, where it was dismantled and sold. The crew and cargo were split between the thirty-first, thirty-second, thirty-sixth and fortieth Nanjing ships, as well as the thirty-third Nanjing ship,[16] and returned to China. In addition to the 53 crew members and 640 crates of copper, the ship had carried 31 *maru* 丸 of sea cucumber and abalone.[17]

In 1706/7, Fukuda Dennoshin 福田傳之進 and other city elders (*machi toshiyori* 町年寄) of Nagasaki received a pheasant from the ninety-first Taiwan ship. As a return gift, "Dennoshin offered thirty *kin* of sea cucumber and thirty *kin* of dried abalone."[18]

15 本船ニ積込申候いりこ・あわひ等、永ゝ日和悪敷候に付、殊外しめり申候。[This ship's load of dried sea cucumber and abalone had, due to many days of poor conditions, become exceedingly damp.] Ibid., 3: 263.

16 Occasionally the number assigned to ships did not strictly conform to the order in which they arrived in port. Therefore, the numbers are not always continuous. The reasons for this irregularity are not clear, but it likely was due to the situation of the ship owner or circumstances in Nagasaki.

17 Ibid., 4: 144, 147–48.

18 Ibid., 4: 152.

FIGURE 6.2 *Dried shark fin.*
COURTESY OF NAGASAKI MUSEUM OF HISTORY AND CULTURE (長崎歴史文化博物館)

In 1707/8 the eighty-first Taiwan ship received permission to acquire supplies of "thirteen *maru* (390 kg) of sea cucumber and dried abalone, and five barrels (*oke* 桶) of soy sauce."[19] On the other hand, in the same month the seventy-sixth Nanjing ship was denied a request to return to China with "seven hundred *kin* of sea cucumber; abalone; dried cuttlefish; shark fin; ceramic ware; rice bowls; mushrooms; all together seven types of goods."[20] Also in the same month, the eighty-first ship had loaded for its return voyage "thirteen *maru* of dried abalone. Yet these did not please the Chinese, and they sought to return them, submitting a request indicating their desire to unload the goods onto land."[21] Finally, in 1707/9 money from the sale of goods from a damaged ship was used to "purchase sea cucumber and abalone, which the Chinese requested to carry with them onto the thirty-ninth [Taiwan] ship."[22]

In 1708/6 a returning cargo vessel requested "these four items: sea cucumber, abalone, dried cuttlefish, and *tengusa* てん草 seaweed," and was given permission to take them on as cargo.[23] In the same month, Cheng Yi Fan 程益凡, owner of the sixty-first ship, received permission "to load the following

19 Ibid., 4: 243.
20 Ibid., 4: 244.
21 Ibid., 4: 246.
22 Ibid., 4: 261.
23 Ibid., 5: 45.

dried goods the next morning: four hundred *kin* of sea cucumber, one hundred *kin* of dried abalone, and two hundred *kin* of dried cuttlefish."[24]

Liu Huaqian 劉華謙, owner of the forty-seventh ship, which returned on 1709/6/19, submitted a request to exchange his cargo of one hundred baskets (*kago* 籠) of sugar for sea cucumber, dried abalone and kelp.[25] In 1709/11, the fifty-sixth Ningbo ship asked to see "samples of dried goods" (*tawaramono tehon* 俵物手本).[26] After being shown kelp, dried cuttlefish, dried sea cucumber, shark fin and dried abalone, they decided to take the goods for their return cargo.

On 1710/9/2, the twenty-eighth Taiwan ship and the thirty-fifth Xiamen 厦門 ship "purchased various goods such as dried cuttlefish, dried abalone and soy sauce."[27] Around 1713/1, the Ningbo and Nanjing ships bought "sea cucumber and abalone" for their return cargo.[28] On 1713/11/27, the returning third Nanjing ship requested to exchange its 120 *maru* of kelp for sea cucumber and dried abalone.[29] On 1714/2/12, the forty-sixth ship requested for its return cargo "dried goods, including two hundred *kin* of sea cucumber, one hundred *kin* of abalone, and one hundred *kin* of shark fin."[30]

2.2 *Dried Sea Cucumber*

The dried sea cucumber loaded as return cargo on Chinese junks was procured from regions surrounding Nagasaki.[31] As discussed below, these goods probably originated in Ōmura 大村 and Hirado 平戸, both domains close to Nagasaki.

Of the various dried goods merchants[32] and "five sea cucumber, dried cuttlefish, shark fin and bonito merchants"[33] active in Nagasaki in 1708/2, two people worked daily in the Chinese quarters, inspecting the goods purchased by Chinese ships for their return cargoes. The eighty-second Nanjing ship, returning on the third day of the sixth month, submitted a request to purchase two *maru* of dried sea cucumber.[34] On 1710/8/26, gifts of dried sea cucumber

24 Ibid., 5: 48–49.

25 砂糖類百かこ餘ロへ牛皮有之候を、何卒いりこ・あわひ・昆布類ニ替させ被下候様。[Would it somehow be possible to exchange our more than one hundred crates (かこ) of sugar, as well as our leather, for sea cucumber, dried abalone, and kelp?] Ibid., 5: 160.

26 Ibid., 5: 198.

27 Ibid., 5: 297.

28 Ibid., 6: 107.

29 Ibid., 6: 210–11.

30 Ibid., 7: 22.

31 Ibid., 2: 387.

32 Ibid., 4: 302.

33 Ibid., 4: 304.

34 Ibid., 5: 30.

FIGURE 6.3 *Dried sea cucumber.*
COURTESY OF NAGASAKI MUSEUM OF HISTORY AND CULTURE (長崎歴史文化博物館)

were distributed to nine ships as thanks for exotic birds sent as presents to Japanese officials.[35] The ships were given a total of 190 *kin* of dried sea cucumber in return for the birds they brought at the request of merchants in Nagasaki,

35 Details for the nine ships are as follows:

Ship	Return gift	Amount	In exchange for
Second Ningbo ship	Sea cucumber	1 box, 15 *kin*	one blue parakeet (*qingyingge* 青鸚哥)
Sixth ship	Sea cucumber	1 box, 25 *kin*	one pheasant (*jinniao* 錦鳥)
Seventh ship	Sea cucumber	1 box, 15 *kin*	one blue parakeet
Ninth Ningbo ship	Sea cucumber	1 box, 15 *kin*	one meadow bunting (*huameiniao* 畫眉鳥)
Eleventh ship	Sea cucumber	2 boxes, 25 *kin*	two pheasants
Eleventh ship	Sea cucumber	1 box, 15 *kin*	one meadow bunting
Twelth Nanjing ship	Sea cucumber	1 box, 15 *kin*	one crested mynah (*bageniao* 八哥鳥)
Fifteenth Nanjing ship	Sea cucumber	1 box, 15 *kin*	one meadow bunting
Sixteenth Nanjing ship	Sea cucumber	1 box, 15 *kin*	four pheasants
Eighteenth Nanjing ship	Sea cucumber	1 box, 25 *kin*	two red-billed wrens (*sōshichō* 相思鳥)

Ibid., 5: 286–87.

which included parakeets, pheasants and mynah birds. On 1712/3/11, the forty-first ship sought dried sea cucumber, shark fin and kelp[36] for its return cargo, but due to disagreements over the price, the trade was unsuccessful.

2.3 *Return Cargo of Chinese Junks during the Hōei* 宝永 *Period (1704–1711)*

Chinese demand for marine products existed from the start of the Edo period. Despite this, marine products represented a smaller portion of the return cargoes of Chinese ships than copper. Not many records remain of the cargoes taken on by Chinese merchant ships for their return journeys after completing their trade in Nagasaki. The only known compilation is the *Tōban kamotsuchō* 唐蠻貨物帳 [Accounts of Trade with China and the Barbarians], held in the Naikaku Bunko collection of the Japanese National Archives.[37] I wish to look at the first account from that compilation, which records the seventh ship of 1709. The record is summarized in table 6.1, and figure 6.4 shows these values in the form of a pie chart:

TABLE 6.1 *Return cargo of the seventh Nanjing ship, 1709*[38]

Goods and Expenditures	Trade value (in *monme* 匁 of silver)	Gold equivalent (in *ryō* 両)	Volume of cargo (in *kin*)
Silver	2,700.000	45.0	
Copper	40,739.060		35,673.4327
Recycled copper goods	1,850.160		1,587.7500
Sundry goods (*komamono iroiro* 小間物色々)	690.000		

36 Ibid., 6: 46.

37 *Tōban kamotsu chō* [Lists of cargoes from China and Europe], parts 1 and 2. Both parts are facsimiles, totaling 1,626 pages. The second part includes Yamawaki Teijiro's 山脇悌二郎 "*Tōban kamotsu chō kaidai*" 唐蠻貨物帳解題 ["Explanatory Notes on *Tōban kamotsu chō*"], 1–12.

38 "寶永六丑年　七番南京船帰帆荷物買渡帳　船頭　沈秋堂　七月十日 [Sixth year of Hōei [1709], Year of the Ox: Record of the Purchase of Return Cargoes for the Seventh Nanjing Ship; Captain Shen Qiutang 沈秋堂. Tenth day of the seventh month]" in *Tōban kamotsu chō*, 1: 2–7.

Goods and Expenditures	Trade value (in *monme* 匁 of silver)	Gold equivalent (in *ryō* 両)	Volume of cargo (in *kin*)
Fox pelts	5,353.000		
Sea cucumber	20,780.970		6,164.0000
Shark fin	1,113.400		475.5000
Dried abalone	5,632.9700		1573.5000
Dried shellfish	70.200		108.0000
Kelp	1,255.815		5,841.0000
Total purchases	77,485.575	1,291.5	
Expenditures	29,614.430	493.5	
Total	**109,800.000**	**1,830.0**	51,423.1827

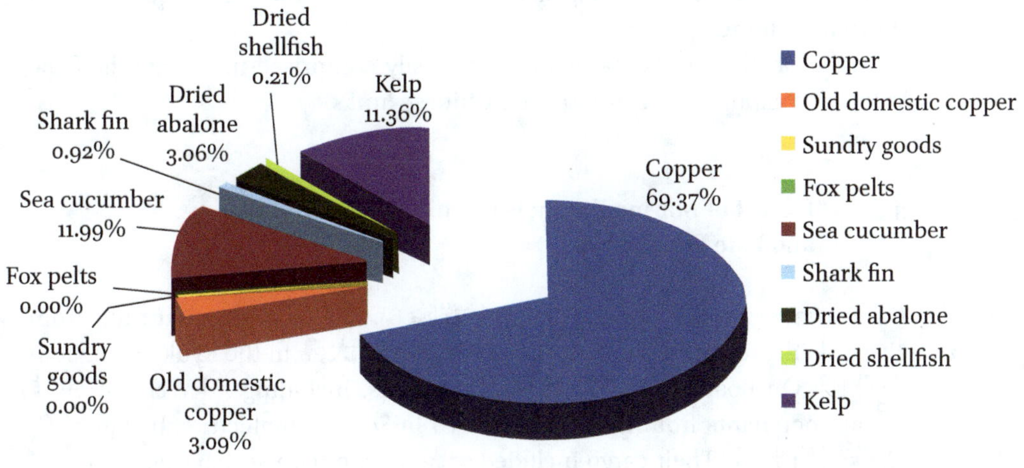

FIGURE 6.4 *Trade value of return cargo of the seventh Nanjing ship, 1709 (by value in silver).*

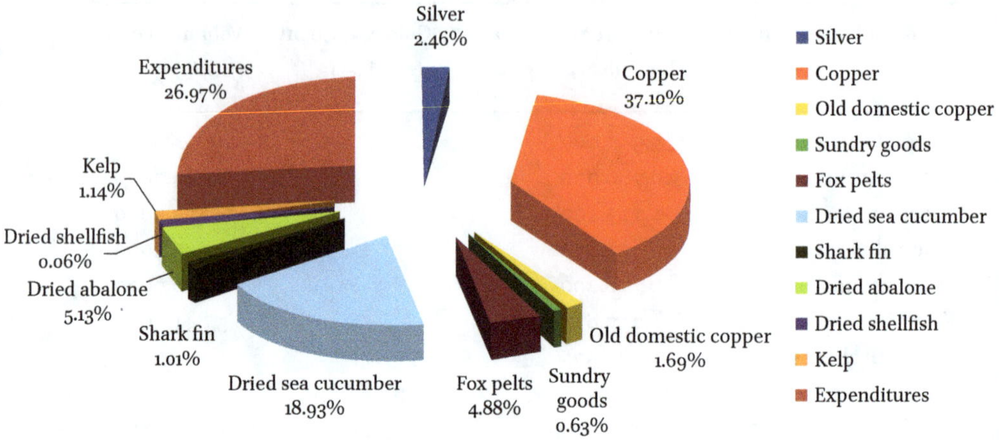

FIGURE 6.5 *Trade value of return cargo of the seventh Nanjing ship, 1709 (by volume).*

In comparison, the following figure 6.5 represents the same goods broken down by volume.

It is clear that in this period copper easily comprised more than half the volume of exports from Nagasaki on Chinese junks.

3 Export of Marine Products to China in the Middle and Late Edo Periods

On 1741/12/8 (Kampō 寛保 1, Qianlong 乾隆 6), a Chinese ship returning from Nagasaki drifted ashore on the island of Ōshima 大島 in the Ryūkyū kingdom 琉球国. On board were fifty-three passengers, including Xu Weihuai 徐惟懐, all merchants from the county of Wu in Suzhou prefecture, Jiangnan 江南蘇州府呉縣. Their cargo included 77,499 *kin* of copper bars, 25,080 *kin* of sea cucumber and abalone, 4,200 *kin* of cockscomb algae (*hong cai* 紅菜) and squid, and 24,000 *kin* of kelp.[39] The proportions of copper, dried goods and kelp were roughly 61.2 percent for copper, 19.8 percent for dried goods, and 19.0 percent for kelp, revealing that the amount of copper was still a large proportion of the total cargo.

An example of a Chinese ship with an even larger cargo of copper may be found in a report by Fujian governor Zhong Yin 鐘音 dated 1756/12/4. Returning to Zhejiang province from Nagasaki, the ship unexpectedly encountered trou-

39 *Rekidai hōan*, 4: 405.

bles at sea and became stranded in the seas near Fuqing 福清 county along the Fuzhou 福州 coast in Fujian. Damaged, the ship was unable to transport its cargo, and so asked the Fujian government to purchase its load of 172,500 *kin* worth of red copper.[40] Thus we see how in 1756, more than 170,000 *kin* of copper was exported from Nagasaki on returning Chinese merchant ships.

In 1762/4, the seventh ship of the previous year returned to China from Nagasaki. I would now like to take a look at its return cargo record. A work titled *Nagasaki hiroku* 長崎秘録 [Secret Records of Nagasaki] lies in the Murakami Collection of the Kariya Municipal Library in Aichi Prefecture. The second volume of that work contains copies of the Nagasaki trade records for the Hōreki years (1751–1764), including information about the return cargoes of three ships from 1762/4 and three ships from 1762/9, out of a total of twelve ships that arrived in Nagasaki in 1761 and fifteen that arrived in 1762.[41] I would like to look at the first record, of the seventh ship from 1762, which returned on 1762/4/22. The following tables summarize the goods purchased:

TABLE 6.2 *1762 Seventh Ningbo ship, value of return cargo*[42]

Item	Value in silver *monme*	Amount in *kin*
Silver	1,100.0	
Copper	118,576.0	102,750.7
Dried sea cucumber	90,586.4	280,201.3

40 浙江歸安縣商人高山輝、從乍浦出口、往販東洋置貨、回棹適遇颶風、於閏九月初五日、飄至閩省福清縣沙塢地方、船隻受傷、莫能駕駛回、帶有紅銅一十七萬二千五百觔、情願就閩收買、以供鼓鋳、・・・。[Gaoshan Hui 高山輝, a merchant from Guian in Zhejiang 浙江歸安縣, set out from Zhapu to acquire goods in Japan. Upon turning his ship, he ran into a storm, and on the fifth day of the intercalary ninth month, washed ashore in Fuqing County, Min [Fujian] Province, in the Shawu region 閩省福清縣沙塢地方. His ship had suffered damage, such that horses were not able to drag it free. He requested that the Min government purchase his cargo of 172,500 *jin* of red copper, which they used to mint coins.] *Gongzhongdang Qianlongchao zouzhe*, 16: 238.

41 *Nagasaki hiroku*, 2.

42 "巳年七番　寧波船買渡荷物帳　宝暦十二年　午四月廿二日帰帆　船頭曹體三　唐人数八拾三人 [Seventh Ningbo Ship: Record of Return Cargo, Returning on the twenty-second day of the fourth month, 1762, Captain: Cao Tisan 曹體三 Crew: 83 Chinese]." *Nagasaki hiroku*, 2: 10–12.

TABLE 6.2 *1762 Seventh Ningbo ship* (cont.)

Item	Value in silver *monme*	Amount in *kin*
Dried abalone	5,688.0	21,040.0
Dried shark fin	1,680.0	840.0
Dried cuttlefish	2,184.0	1,680.0
Cockscomb algae	6,300.0	6,000.0
Kelp	53,126.4	159,060.0
Sake	275.0	
Soy sauce	40.0	
Gold-leaf lacquer ware	330.0	
Expenditures in Nagasaki	50,570.9	
Total	330,456.7	

TABLE 6.3 *1761–1762 Chinese ships, return cargoes; volume of copper and marine products* (*in* kin)

Year	1761	1762	1762	1762	1762	1762	1762
Ship	Third Xiamen	Seventh Ningbo	Eighth Nanjing	Ninth Ningbo	Tenth Ningbo	Twelth Nanjing	Sixth Ningbo
Captain	Cui Jingsan 崔景三	Cao Tisan 曹體三	Gong Zixing 龔子興	Wang Shengwu 汪繩武	Yuan Shenghuan 袁盛寰	Cheng Yutian 程玉田 & Zhao Zhuruo 趙主若	Fang Gongchen 方拱宸 & Shen Lunxi 沈綸溪
Crew	83	83	80	115	75	112	77
Copper	100,000	102,750.7	150,000	142,763.5	100,000	161,091.3	132,816
Sea cucumber	27,752.4	28,021.3	26,390.1	39,782.9	22,829	24,266.2	22,920
Abalone	2,258.9	2,040	2,123.3	7,836.6	5,181.8	3,840	5,038.7
Shark fin	1,080	840	1,080	1,080	420	420	600
Kelp	155,608.7	159,060.8	136,139.4	157,586.4	179,097.8	158,377.6	165,316.4
Leaf No.	7–8	10–11	17–18	21–22	25	28–29	34–35

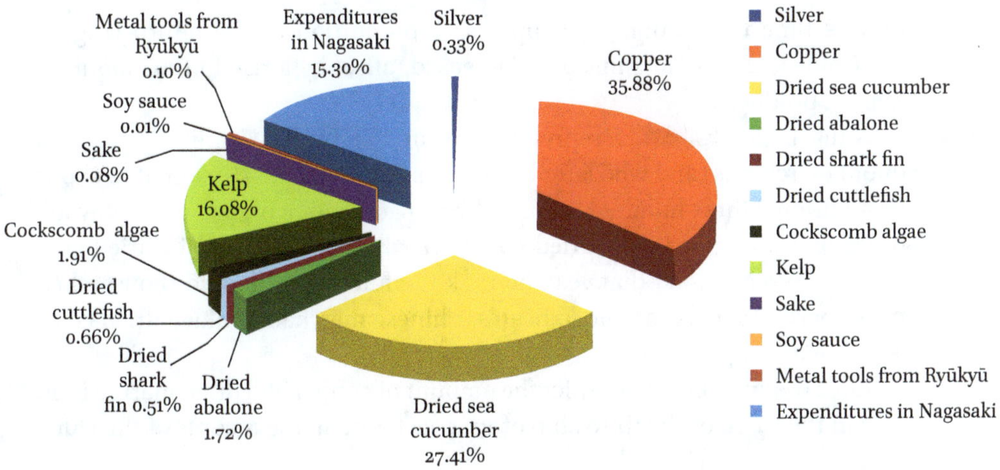

FIGURE 6.6 *1762 seventh Ningbo ship, value of return cargo (by value).*

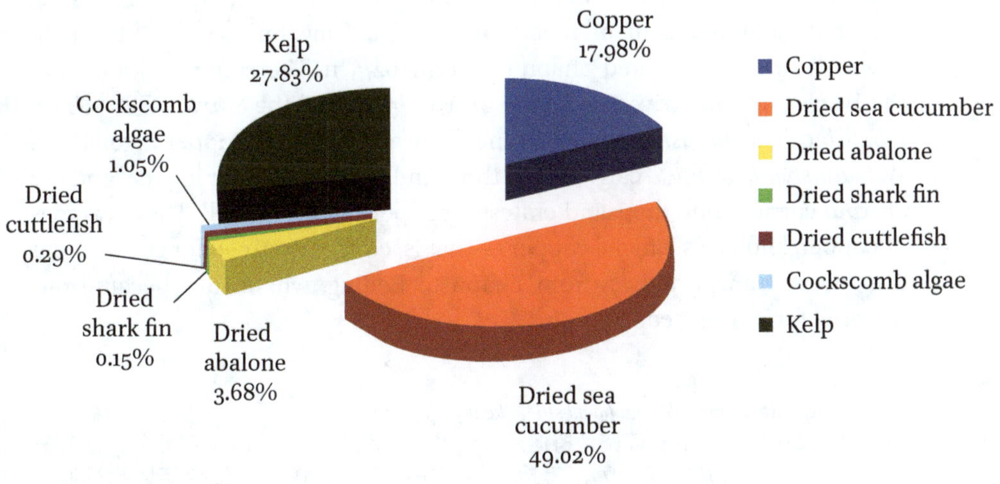

FIGURE 6.7 *1762 seventh Ningbo ship, value of return cargo (by volume).*

By this time the amount of copper as a proportion of the return cargo of the Chinese merchant ships had decreased, offset by a rise in the amount of marine products.

Wang Lüjie 王履階, a representative of the Twelve Families 十二家 of quota merchants (and who later became a state merchant),[43] sought to export an additional three hundred thousand *kin* of goods, of which 70 percent would be copper and 30 percent dried goods. Yet due to a shortage of dried good imports into Nagasaki that year, he took the full three hundred thousand *kin* in copper.[44] As this example indicates, Chinese merchants valued dried goods as much as copper.

Table 6.4 and figure 6.8 depict the amount of copper and dried marine products in the cargo of the third ship of 1764, as found in the records of the Dutch Trading House in Nagasaki.[45]

Accounts of how much copper was loaded onto returning Chinese vessels appear in the palace memorials of Qing court officials. The first account, from Zhejiang governor Xiong Xuepeng 熊学鵬, dates from 1763/1/24: Under the command of merchant-official Fan Qingzhu 范清注[46] the ship *Yangshihe* 楊士合 entered Zhapu port on 1762/5/11. Thereafter we know that the *Weiyuansheng* arrived in Zhapu on 1762/5/17, and the *Shixinli* 施新利 on 1762/9/30. The *Yangshihe* carried "one thousand crates of copper ingots;" the *Weiyuansheng* 魏元盛, carried "one thousand crates of copper ingots;" and the *Shixinli* carried "one thousand crates of copper ingots" as well. These cargoes of copper, known as 'official copper'—that is, copper destined for government use—were transported on from Zhapu to Beijing, where they became raw material for minting copper coins.[47]

---

43 Matsuura, *Shindai kaigai bōeki shi no kenkyū*, 151–53.

44 唐國ヨリ元絲銀三百貫目持渡ル。此代リ銅三拾萬斤、内正銅七分、俵物三分、可被相渡約條ニテ、二拾ケ年可持渡憑文渡置ル。但今年俵物拂底ニ付、正銅三十萬斤相渡サル。[[Wang Lüjie] carried three hundred thousand *monme* of silver currency from China. In exchange for this he sought goods valued at three hundred thousand *kin* of copper, 70 percent actual copper and 30 percent dried goods. He hoped to set up a contract for these conditions lasting twenty years. However due to a shortage of dried goods this year, he received 300,000 *kin* of copper.] *Nagasaki jitsuroku taisei* 11, 278, entry for Hōreki 13.

45 Nagazumi, *Tōsen yushutsunyūhin sūryō ichiran* 1637–1833 *nen*, 265.

46 Matsuura, *Shindai kaigai bōeki shi no kenkyū*, 148.

47 官商范清注、採辦銅斤、由乍（乍浦）起運、鮮供蘇省、分鮮五省官銅、該官商、乾隆二十七年、辦回船戶楊士合船、銅斤一千箱、於五月十一日、由乍進口、入境即於十二日、起運赴蘇、於本月二十三日出境、又船戶魏元盛船、銅斤一千箱、於閏五月十七日、由乍進口、入境即於二十六日、起運五

TABLE 6.4 *Volume of copper and dried marine products, return cargo of the third ship, 1764 (in* kin)

1764 Third ship	Copper rods	Dried sea cucumber	Dried abalone	Shark fin	Kelp	Dried cuttlefish
	172,550	20,000	6,360	540	174,545	1,080

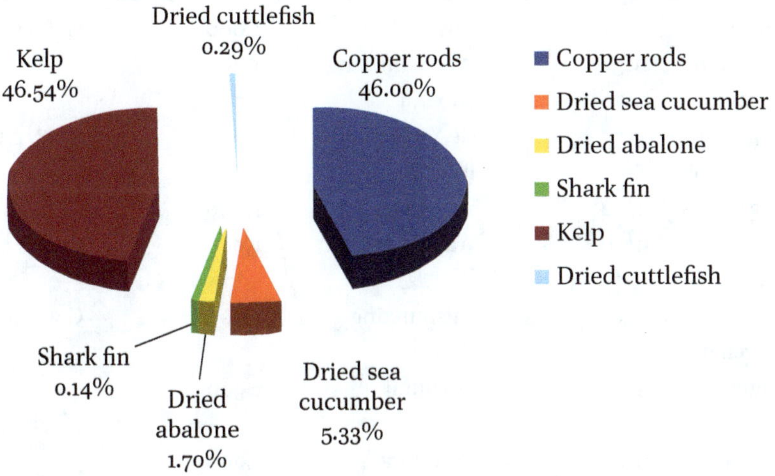

FIGURE 6.8 *Volume of copper and dried marine products, return cargo of the third ship, 1764.*

百四十箱、又於二十九日、起運四百六十箱赴蘇、俱於六月初一日出境、又船戸施新利船、銅斤一千箱、於九月三十日、由乍進口、入境即於十月初六、七等日、起運赴蘇、於十月十三日出境、均経沿途各県、加謹稽査防護、並無偸盜·沈溺事。[Merchant-official Fan Qingzhu obtained copper ingots. He set off from Zhapu and supplied Suzhou and five other provincial governments with copper. Under that merchant-official [Fan], the ship *Yangshihe*, carrying one thousand crates of copper ingots, left Zhapu on the eleventh day of the fifth month. It entered our territory on the twelfth, and left again for Suzhou on the twenty-third of the same month. The

Similar records exist through 1783, and, as shown in table 6.5, the majority of trade vessels returning to Zhapu from Nagasaki were laden with one thousand crates of copper.[48]

TABLE 6.5 *Chinese vessels under the command of Qing merchant-officials returning to Zhapu*

Western year	Chinese year	Date of entry to Zhapu	Name of vessel	Copper cargo (crates)	Source
1762	Qianlong 27	5/11	Yangshihe	1,000	G 28/1/24
	State merchant:	IC5/17	Weiyuansheng	1,000	
	Fan Qingzhu	9/30	Shixinli	1,000	
1763	Qianlong 28	3/7	He Yanbao 何延寶	1,000	G 28/1/6
	State merchant:				
	Fan Qingzhu	4/12	Yangshihe	1,000	
		9/10	Wangyongqing 王永慶	1,000	
1764	Qianlong 29	4/1	Shishunxing 史順興	1,000	G 30/1/22
	State merchant:				
	Fan Qinghong 范清洪	5/13	Jinyuntai 金允泰	900	
		9/27	Yangshihe	1,000	
		11/15	He Tingbao 何廷寶	1,000	

Weiyuansheng, carrying one thousand crates of copper ingots, left Zhapu on the seventeenth day of the intercalary fifth month. It entered our territory on the twenty-sixth. It initially transported 540 crates, and then on the twenty-ninth transported another 460 crates to Suzhou, leaving our territory on the first day of the sixth month. The *Shixinli*, with one thousand crates of copper, left Zhapu on the thirtieth day of the ninth month. It entered our territory on the sixth or seventh day of the tenth month. It travelled onwards toward Suzhou, leaving our borders on the thirteenth day of the tenth month. It stopped at each prefecture along the way, and by taking great care and precautions, avoided both piracy and disaster.] *Gongzhongdang Qianlongchao zouzhe*, 16: 655–57.

48 Matsuura, *Shindai kaigai bōeki shi no kenkyū*, 354–56.

Western year	Chinese year	Date of entry to Zhapu	Name of vessel	Copper cargo (crates)	Source
1765	Qianlong 30	3/6	Weiyuansheng	1,000	Z 31/1/7
	State merchant:	9/4	Linyongshun 林永順	500	
	Fan Qingji 范清濟		Wang Yongsheng* 王永順	721	
		9/8	Wan Youshun 萬友順	1,000	
		10/9	Weiyuansheng	1,000	
			Linyongshun*	500	
1766	Qianlong 31	2/12	Jinyuntai	1,000	Z 32/1/4
	State merchant:	2/24	Shishunxing	1,000	
	Fan Qingji	5/12	Linyongshun	1,000	
		5/14	Weiyuansheng	1,000	
		9/24	He Tingbao	1,000	
		9/27	Yangshihe	1,000	
		10/1	Zhoushunxing 周順興	1,000	
		10/15	Shishunxing	1,000	
		10/19	Weiyuansheng	1,000	
1767	Qianlong 32	5/18	Jinyuntai	1,000	G 33/1/3
	State merchant:	10/6	Xuxianchun 許咸春	1,000	
	Fan Qingji	10/24	Jinwanyu 金萬裕	1,000	
		11/20	Wanrixin 萬日新	1,000	
		11/20	Weiyuansheng	1,000	
1768	Qianlong 33	9/3	Wanrixin	1,000	Z 34/1/27
	State merchant:	9/3	Linyongshun	1,000	
	Fan Qingji	9/3	He Tingbao	1,000	
		11/3	Jinwanyu	1,000	
		11/13	Xuxianchun	1,000	
1769	Qianlong 34	3/18	Yangshihe	1,000	Z 35/1/7
	State merchant:	9/20	Shishunxing	1,000	
	Fan Qingji	9/26	Weiyuansheng	1,065	
		11/6	Jinwanyu	1,000	
		11/21	He Tingbao	1,000	

TABLE 6.5 *Chinese vessels under the command* (cont.)

Western year	Chinese year	Date of entry to Zhapu	Name of vessel	Copper cargo (crates)	Source
1770	Qianlong 35	5/5	Wanrixin	1,000	Z 36/1/11
	State merchant:	5/7	Xuxianchun	1,000	
	Fan Qingji	9/19	Yangshihe	1,000	
		9/22	Linyongshun	1,000	
		9/22	Jinwanyu	1,000	
		10/8	He Tingbao	1,000	
1771	Qianlong 36	3/13	Wanrixin	1,000	Z 36/12/11
	State merchant:	3/13	Shishunxing	1,000	
	Fan Qingji	9/9	Jinwanyu	1,000	
		9/10	Zhouyong'an 周永安	1,000	
		9/10	Yueshunxing 岳順興	1,000	
		9/10	Fanjizong 范繼宗	1,000	
1772	Qianlong 37	2/18	Linyongshun	1,000	Z 38/1/6
	State merchant:	4/22	He Tingbao	1,000	
	Fan Qingji	9/8	Weiyuansheng	1,000	
		9/19	Shishunxing	1,000	
		9/19	Wanrixin	1,000	
1773	Qianlong 38	3/19	Zhouyong'an	1,000	G 38/12/1
	State merchant:	3/22	Jinwanyu	1,000	
	Fan Qingji	8/21	Linyongshun	1,000	
		10/11	Jiangxiangtai 江祥泰	1,000	
		10/18	He Tingbao	1,000	
		10/18	Wanrixin	1,000	
1774	Qianlong 39	3/12	Weiyuansheng	1,000	G 39/12/6
	State merchant:	3/25	Shishunxing	1,000	
	Fan Qingji	3/26	Fanjizong	1,000	
		3/26	Jinyuanbao 金源寶	1,000	
		4/27	Zhoushunli 周順利	1,000	
		9/18	He Tingbao	1,000	
		9/30	Linyongshun	1,000	

Western year	Chinese year	Date of entry to Zhapu	Name of vessel	Copper cargo (crates)	Source
1775	Qianlong 40	2/2	Wanrixin	1,000	Z 40/12/27
	State merchant:	2/2	Weiyuansheng	1,000	
	Fan Qingji	4/5	Hongchengtai 洪成泰	1,000	
		5/6	Fanjizong	1,000	
		5/9	Jinyuanbao	1,000	
		9/28	Zhoushunli	1,000	
		10/24	He Tingbao	1,000	
		Int 10/8	Jinyuanbao	1,000	
1776	Qianlong 41	5/1	Linyongshun	1,000	Z 41/12/12
	State merchant:	8/22	Weiyuansheng	1,000	
	Fan Qingji	9/19	Hongchengtai	1,000	
		9/19	Fanjizong	1,000	
		9/21	Wanrixin	1,000	
		10/11	Jinyuanbao	1,000	
1777	Qianlong 42	4/16	Shishunxing	1,000	G 42/11/24
		9/8	Zhoushunli	1,000	
		9/11	Wanrixin	1,000	
		10/3	Weiyuansheng	1,000	
		10/9	Linyongshun	1,000	
		10/9	He Tingbao	1,000	
1778	Qianlong 43	4/20	Hongchengtai	1,000	G 43/12/4
	State merchant:	5/6	Jinyuanbao	1,000	
	Fan Qingji	5/23	Wanrixin	1,000	
		9/27	Shishunxing	1,000	
		10/6	Fanjizong	1,000	
		10/8	Zhoushunli	1,000	
		10/8	Weiyuansheng	1,000	
1779	Qianlong 44	3/9	He Tingbao	1,000	Z 44/12/10
	State merchant:	4/7	Linyongshun	1,000	
	Fan Qingji	4/7	Fanjizong	1,000	
		4/7	Jinyuanbao	1,000	
		10/12	Zhouwanshun 周萬順	1,000	
		10/17	Wanrixin	1,000	
		11/4	Shishunxing	1,000	

TABLE 6.5 *Chinese vessels under the command* (cont.)

Western year	Chinese year	Date of entry to Zhapu	Name of vessel	Copper cargo (crates)	Source
1780	Qianlong 45	4/22	Weihongsheng 魏宏勝	1,000	Z 45/11/28
	State merchant:				
	Fan Qingji	4/24	Fanjizong	1,000	
		5/8	Jinyuanbao	1,000	
		5/18	He Tingbao	1,000	
		10/12	Linyongshun	1,000	
		10/15	Zhouwanshun	1,000	
		10/18	Wanrixin	1,000	
1781	Qianlong 46 State merchant: Fan Qingji	262,533 *kin* of 'foreign copper': "A total of 262,533 *jin* of foreign copper left Suzhou on the seventh day of the sixth month of Qianlong 46."[49]			G 46/9/19
1782	Qianlong 47	5/9	Zhouwanshun	1,000	G 47/11/27
	State merchant:	5/10	Weihongsheng	1,000	
	Fan Qingji	5/12	He Tingbao	1,000	
		10/19	Jinyuanbao	1,000	
		10/19	Wanrixin	1,000	
		10/25	Fanchangji 范常吉	1,000	
1783	Qianlong 48	4/12	Linyongshun (Fan)	1,000	G 48/12/5
	State merchants: Fan Qingji Wang Shirong 王世榮	11/1	Wanrixin (Wang)	1,000	

'Z' refers to the *Zhupi zouzhe* 硃批奏摺 in the Chinese Number One Historical Archives 中国第一歴史檔案館, while 'G' refers to the *Gongzhongdang Qianlongchao zouzhe* in the Taiwan Palace Museum. Arabic numerals denote the dates of memorials, and are given in the old calendar.
[*denotes ships that returned together with the preceding ship]

49 *Gongzhongdang Qianlongchao zouzhe*, 48: 837.

As the above table illustrates, each of the Chinese merchant ship sent to Nagasaki and commanded by merchant-officials returned with one thousand crates—one hundred thousand *kin*—of copper.

4 The Export of Marine Products to China in 1803 and 1804

According to the *Nagasaki kaisho gosatsumono* 長崎會所五冊物 [Five-volume Record of the Nagasaki Trade Office],[50] which details the accounts of the Nagasaki clearing house for foreign trade accounts following the revised trade laws of 1791, "those responsible for selling copper to the Chinese ships sold a predetermined amount equivalent to 110,000 *monme* of silver."[51] Each ship "took on one hundred thousand *kin* of copper"[52] carried in one thousand boxes of one hundred *kin* each.[53]

The *Nagasaki kaisho gosatsumono* also mentions dried goods. In this period, when only ten ships were allowed to trade at Nagasaki every year, each Chinese vessel exported roughly 95,400 *monme* worth of dried goods, meaning the total volume reached 954,000 *monme* per year. Dried goods constituted dried sea cucumber, dried abalone and shark fin.[54]

The *Nagasaki kaisho gosatsumono* categorized all goods aside from copper and dried goods as miscellany (*shoshiki* 諸色). Such miscellany included not only kelp, which comprised a large chunk of the return cargoes of Chinese vessels, but also dried seaweed products such as *tengusa* seaweed, and other dried marine products such as dried cuttlefish, dried shrimp and scallops.[55]

---

50 *Nagasaki kaisho gosatsumono*, 1–243.

51 Ibid., 39.

52 Ibid., 40.

53 唐人渡方之節ハ、壹箱宛正味百斤ニ掛改. [The regulations were revised so that each crate of the Chinese merchants contains 100 *kin*.] Ibid., 39. Also see Matsuura, "Shindai kanshō saiben yōdō benkai senseki", 38–45.

54 *Nagasaki kaisho gosatsumono*, 40.

55 昆布・鯣・茯苓・鶏冠草・所天草・鰹節・千切レ砂食煎海鼠・千切レ蟲入干鮑・藤海鼠・刻昆布・干海老・五倍子・椎茸・干瀬貝・寒天・樟脳・いたら貝・獺皮・御種人参・會津・雲州和人参・銅器物・流金道具・蒔繪小間物・呉服、其外樽物等を諸色と唱、唐人共買渡申候、則壹船當時買渡平均六拾三貫六百目程之積り、壹ケ年拾艘六百三拾六貫目程、出帆之時々荷造仕、唐人相好候品々買渡候儀ニ御座候。[Kelp, cuttlefish, *fu ling* mushrooms, cockscomb algae, *tengusa* seaweed, dried bonito flakes, diced dried sea cucumber, diced dried abalone speckled with holes 千切レ蟲入干鮑, black sea cucumber, chopped kelp 刻昆布, dried shrimp, sumac gallnuts, *shiitake* mushrooms, dried sea snail 干瀬貝,

TABLE 6.6 *Value in silver of return cargoes for Chinese vessels, 1803–1804 (in* kanme)[56]

Year	Ship No./Origin	Ship Name	Trader	Copper
1803	Seventh Nanjing*	Yongxing 永興	Twelve Families	115,00
1803	Eighth Ningbo*	Rixin 日新	Wang	115,00
1803	Tenth Nanjing*	Jieji 皆吉	Twelve Families	115,00
1803	First Nanjing	Jinquansheng 金全勝	Wang	115,00
1803	Second Nanjing	Yongtai 永泰	Twelve Families	115,00
1803	Fourth Ningbo	Yuansheng 源盛	Wang	115,00
1803	Third Ningbo	Yongbao 永寶	Twelve Families	115,00
1803	Fifth Ningbo	Yuanbao 源寶	Wang	115,00
1803	Seventh Nanjing	Dawan'an 大萬安	Wang	115,00
1803	Sixth Nanjing	Jieji	Twelve Families	115,00
1804	Eighth Ningbo*	Yongxiang 永祥	Twelve Families	115,00
1804	Ninth Ningbo*	Jinquansheng	Wang	115,00
1804	Tenth Nanjing*		Twelve Families	115,00
1804	Third Nanjing	Desheng 得勝	Wang	115,00
1804	Unnumbered*	Yuansheng	Wang	115,00
1804	First Ningbo	Yongtai	Twelve Families	115,00
1804	Second Ningbo	Dawan'an	Wang	115,00
1804	Fourth Ningbo	Rixin	Wang	115,00
1804	Seventh Nanjing	Yongxing		115,00
1804	Unnumbered	Yongbao	Twelve Families	115,00
Totals				2,300,00
Averages				115,00

[* indicates ships that left for Nagasaki the previous year; hence their numbers are not aligned with the year they returned to China]

seaweed-based gelatin, camphor, scallops, otter pelts, Korean ginseng 御種人参, Japanese ginseng from Aizu and Izumo, copperware, gold-leaf lacquer ware 流金道具, lacquer goods, kimono fabrics, and all other barreled goods are called miscellany. These goods were purchased by the Chinese, and it was stipulated that one ship would purchase an average of 63,600 *monme* of goods. Thus the ten ships per year carried a total of 636,000 *monme*. At the time of the ships' departure, both wholesalers and Chinese merchants were happily engaged in the trade of goods.] Ibid., 41.

56 This data was originally recorded in units of *kanme*, rather than the more common *monme*. 1 *kanme* is equal to 1,000 *monme*.

oods	Misc	Kelp	Total Cargo Value	Source (leaf no.)
77	60,785		266,962	8
55	59,537		264,091	18
60	67,840		284,600	27
48	65,365		278,413	38
70	66,913		282,283	48
60	67,040		282,600	58
54	57,770		259,424	87–88
47	60,164		265,411	100–01
78	62,119		270,297	113
60	89,440		338,600	124–25
14	71,609		294,023	23
84	87,523		333,807	35
42	83,028		322,570	45
19	65,613		279,032	55
18	58,218	25,300	256,736	65
26	60,350		265,876	89
81	59,587		263,968	99
90	65,126		277,816	109
24	65,350		278,374	137
50	59,350	25,300	259,000	141
57	1,332,727		5,623,883	
23	66,636		281,194	

The *Kyōwa sannen idoshichū shuppan hikiawasechō* 享和三年亥年中出帆引合帳,[57] and the *Bunka gannen nedoshichū shuppan hikiawasechō* 文化元子年中出帆引合帳,[58] both in the collection of the Nagasaki Museum of History and Culture, contain records and calculations of the value of return cargoes of Chinese ships arriving in Nagasaki in 1802, as well as records of the expenditures of their crews while in Nagasaki. The first record bears the title "The Twelve Families. Index of the Original Values in Silver of the Seventh Nanjing Ship, Year of the Dog: the *Yongxing* 十二家戌七番南京船元代銀引目録 永興." The 'Twelve Families' were a group of Chinese traders, a private [i.e. not court-affiliated] group of merchants. The name Wang 王, which will appear below, denotes a state merchant designated by the Qing court.[59] *Yongxing* was the name of a Chinese vessel.[60] This record shows that, out of a total trade value of 266,962 *monme* of the *Yongxing*, 115,000 *monme* were spent on copper, slightly more than 91,177 *monme* on dried goods, and slightly more than 60,784 *monme* on miscellaneous items. Proportionally copper represents 43 percent, dried goods 34 percent and miscellaneous items 23 percent.[61]

If we compare these figures to the twenty ships of 1803 and 1804 (ten per year), looking at the average total cargo values and proportions, we see little change: copper averaged 115,000 *monme*, bagged goods just over 97,000 *kin*, and miscellany slightly more than 66,000 *kin*. Proportionally speaking copper was on average 41 percent, bagged goods 35 percent and miscellany 24 percent of the total cargo. We may imagine that the percentages by value for return cargoes throughout this period are similar.

---

57 *Kyōwa sannen idoshichū shuppan hikiawasechō*.

58 *Bunka gannen nedoshichū shuppan hikiawasechō*.

59 Matsuura, *Shindai kaigai bōeki shi no kenkyū*, 144–46.

60 Ibid., 289.

61 The value of the seventh ship's cargo and a breakdown of its contents—copper, dried goods and miscellany—is as follows:
A total of 266,962.3587 *monme* was used to purchase goods:
- 115,000 *monme* for 100,000 *kin* of copper (100 *kin* cost 115 *monme*)
- 151,962.3587 *monme*
  - 91,177.4152 *monme* of dried goods
  - 60,784.9435 *monme* of miscellany

See *Kyōwa sannen idoshichū shuppan hikiawasechō*, leaf 8.

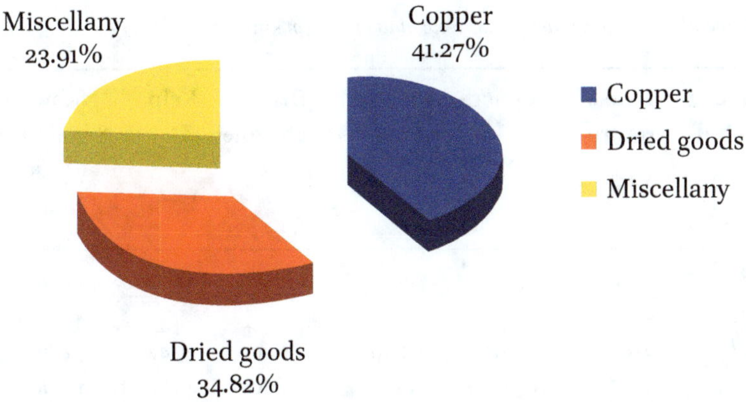

FIGURE 6.9 *Value in silver of return cargoes for Chinese vessels, 1803–1804 (in* monme).

5 Export of Marine Products to China in the Bakumatsu Period

The records of the Dutch Trading House in Nagasaki include the amounts of return cargo on Chinese ships in 1833.[62] These data appear in the following table.

Figure 6.10 represents the average return cargoes of copper, sea cucumber, dried abalone, kelp and gelatin from the fifth through ninth ship. It immediately becomes clear that the quantity of kelp was quadruple that of copper, indicating that from the 1830s onward kelp exports increased dramatically.

6 Cargo Sent to Shanghai on the Official Bakufu Ship Senzaimaru 千歳丸, 1862

After dismantling its policy of national seclusion in the 1850s, the Tokugawa *bakufu* sent its first official ship overseas. The *bakufu* relied on the history of the return cargoes of Chinese vessels calling at Nagasaki to determine what cargo to take on board this ship, the *Senzaimaru*.[63] The chosen goods are shown in the following table.

---

62 Nagazumi *Tōsen yushutsunyūhin sūryō ichiran 1637–1833nen*, 328.

63 Kawashima, "Saisho ni kokoromita Shanhai bōeki," 115–66.

TABLE 6.7 *Principle cargo of Chinese ships returning in 1833 (in* kin*)*[64]

Ship number	Date of arrival	Date of return	Copper	Sea cucumber	Dried abalone	Kelp	Seaweed gelatin (寒天 *kanten*)
5	1832 12/1	1833 4/6	105,000	20,000	25,000	490,000	12,000
6	12/4	4/6	105,000	21,000	25,000	394,000	12,000
7	12/4	4/6	105,000	21,000	25,000	442,000	12,000
8	12/6	4/6	105,000	21,000	25,000	418,000	12,000
9	1833 1/18	4/6	105,000	21,000	25,000	406,000	12,000
Average			105,000	20,800	25,000	430,000	12,000

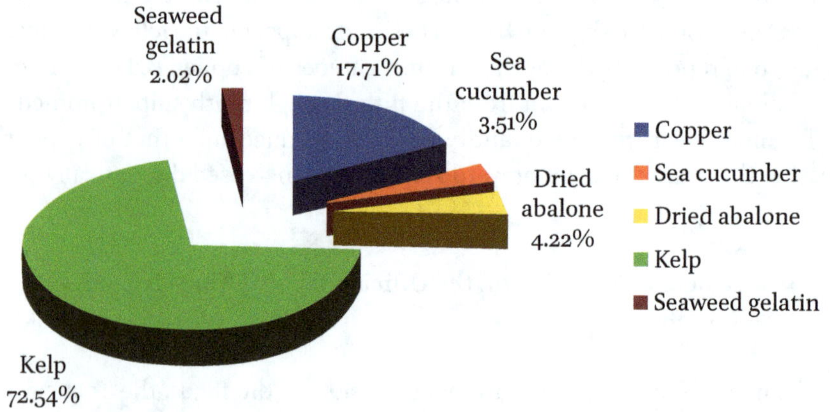

FIGURE 6.10 *Average cargo volume of five Chinese ships returning in 1833.*

64 Dates of arrival and return are taken from the *Tōsen shinkō kaitōroku Shimabara bon Tōjin fūsetsu gaki wappu tome*, 208–9; figures for cargo are taken from Nagazumi, *Tōsen yushutsunyūhin sūryō ichiran* 1637–1833 *nen*, 328.

TABLE 6.8 *Cargo on the ship* Senzaimaru *in 1862*

Goods from Nagasaki clearing house for foreign trade accounts	
Sea cucumber	24,000 *kin*
Dried abalone	36,000 *kin*
Shark fin	1,800 *kin*
Tengusa seaweed	12,000 *kin*
Cockscomb seaweed	3,000 *kin*
Mitsuishi-kelp 三ツ石昆布[65]	3,600 *kin*
Goods from Nagasaki Market	
First grade dried cuttlefish	3,000 *kin*
Second grade dried cuttlefish	2,550 *kin*
Windmill palm leaf sheathes	5,687 *kin*
Japanese ginseng from Aizu	2,500 *kin*
Japanese ginseng from Izumo	2,500 *kin*
Coal[66]	250,000 *kin*
Goods from individual merchants	
Gelatin from *tengusa* seaweed	1,800 *kin*
Mitsuishi kelp[67]	26,424 *kin*

The above items were those carried in large quantities; an additional sixty or so goods were transported in lesser amounts. Although the cargoes of earlier Chinese junks returning from Nagasaki were ostensibly used as a reference, no copper was included in the *bakufu* ship's cargo. A new product replaced it: coal.

As Japan's domestic copper production declined over the course of the Edo period, so too did copper exports. By the time the Tokugawa *bakufu* dispatched the *Senzaimaru*, the system of copper exports was already collapsing. Anxious to dispatch the *Senzaimaru* to Shanghai, the *bakufu* selected coal from the Takashima Mine near Nagasaki to serve as the ship's ballast. In the last years of Tokugawa rule, Saga Domain and the Scottish merchant Thomas Blake Glover (1838–1911) excavated the Takashima Mine. Glover had travelled to Shanghai in 1859, where he joined Jardine Matheson Holdings Ltd. (known as *Yihe*

65 Ibid., 124–25.

66 Ibid., 133–34, 138–39.

67 Ibid., 139.

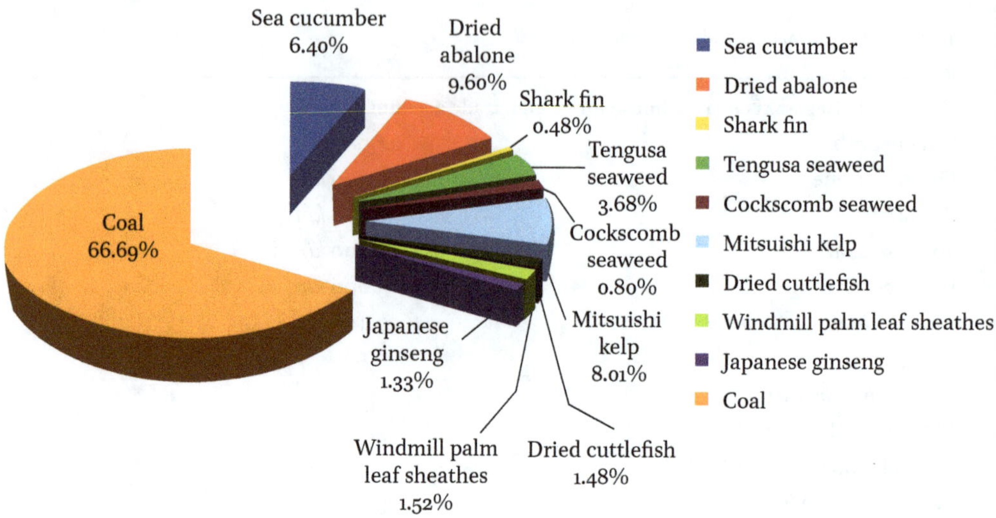

FIGURE 6.11 *Cargo on the ship* Senzaimaru *in 1862.*

yanghang 怡和洋行 in Chinese). In 1859/9, he arrived in Nagasaki, which the *bakufu* had just opened as a treaty port. Glover became Jardine Matheson's representative in Nagasaki, where he also started his own trading company, thus entering the trade industry. He probably helped mediate an increase in shipping at Shanghai: the coal needed to fuel those ships was not available near Shanghai, giving Glover an opportunity to make a profit by shipping coal from Japan. Thereafter foreign ships carried the coal trade between Nagasaki and Shanghai.

7 Sources of Japanese Marine Products

Records of the original production areas of the various dried marine products discussed above appear in the "Tawaramono shoshiki detokoro 俵物諸色出所 [Origins of Miscellaneous and Dried Goods]," contained in the *Kaban kōeki meisaiki* 華蠻交易明細記, which details the various places in Japan whence goods were imported to Nagasaki.[68] These accounts date from the mid-18th century to the early 19th century, and allow us to consider the production areas

68 *Kaban kōeki meisaiki*, 245–425.

of each marine product in the mid to late Edo period.[69] Dried sea cucumber was produced in southern Hokkaido and wide areas of coastal Honshū. In particular, over fifty percent of sea cucumber was produced in present-day Hokkaido, Aomori, Kōchi, Yamaguchi and Ishikawa prefectures. Nearly 80 percent of dried abalone came from the Hirado, Gotō and Tsushima regions of present-day Nagasaki Prefecture. During the Edo period, coastal shipping transported these goods to Nagasaki, whence they were exported to China. The only origin of kelp recorded in the *Kaban kōeki meisaiki* is Matsumae, with "roughly 790,000 *kin*" produced,[70] suggesting that most kelp was produced in what is today southern Hokkaidō before being imported to Nagasaki via Shimonoseki on cargo ships plying the Sea of Japan coast.

TABLE 6.9 *Geographical origins of sea cucumber* (*Units*: kin)[71]

Region	Place	Quantity
Present-day Hokkaidō region	Matsumae 松前	62,100
Present-day Tōhoku region	Tsugaru 津軽	33,800
	Nanbu 南部	3,070
	Sendai 仙台	2,800
Present-day Kantō region	Awa 安房	3,100
	Kazusa 上総	1,060
	Musashi 武蔵	4,800
	Sōshū 相州	8,500
	Owari 尾張	1,200
	Mikawa 三河	11,000

69 Ogawa Kuniharu 小川国治 demonstrates how the Edo *bakufu* and individual domains controlled the production of dried goods through the Edo period in his *Edo bakufu yushutsu kaisanbutsu no kenkyū—tawaramono no seisan to shūka kikō* [*Studies on Marine Products Exported by the Edo Bakufu: Structures of Production and Collection of Dried Goods*]. For an account that draws upon Dutch documents to explain the state of marine products sent from Japan to China, see Arai, *Kinsei kaisanbutsu bōeki shi no kenkyū—Chūgokumuke yushutsu bōeki to kaisanbutsu*. As for the state of exports to China, Ogawa and Arai both refer to Matsuura, "Nisshin bōeki ni yoru tawaramono no Chūgoku ryūnyū ni tsuite": Ogawa for content concerning the Meiji period onward, and Arai for information regarding the Edo period.

70 *Kaban kōeki meisaiki*, 389.

71 Ibid., 388–89.

TABLE 6.9 *Geographical origins* (cont.)

Region	Place	Quantity
Present-day Hokuriku region	Noto 能登	18,800
	Echizen 越前	1,100
	Etchū 越中	450
Present-day Kinki region	Ise 伊勢	11,000
	Shima 志摩	7,100
	Harima 播磨	2,000
	Awaji 淡路	2,000
	Tango 丹後	1,330
	Kishū 紀州	850
Present-day Shikoku region	Awa 阿波	2,500
	Sanuki 讃岐	2,320
Present-day Chūgoku region	Aki 安芸	28,000
	Suō 周防	23,000
	Bizen 備前	15,000
	Nagato 長門	11,250
	Izumo 出雲	2,800
	Bitchū 備中	700
	Iwami 石見	500
Present-day Kyūshū region	Ōmura 大村	14,120
	Hirado 平戸	12,100
	Tsushima 対馬	8,580
	Karatsu 唐津	6,300
	Amakusa 天草	5,700
	Iki 壱岐	3,000
	Higo 肥後	2,400
	Satsuma 薩摩	1,200
	Chikuzen 筑前	320
	Bungo 豊後	270

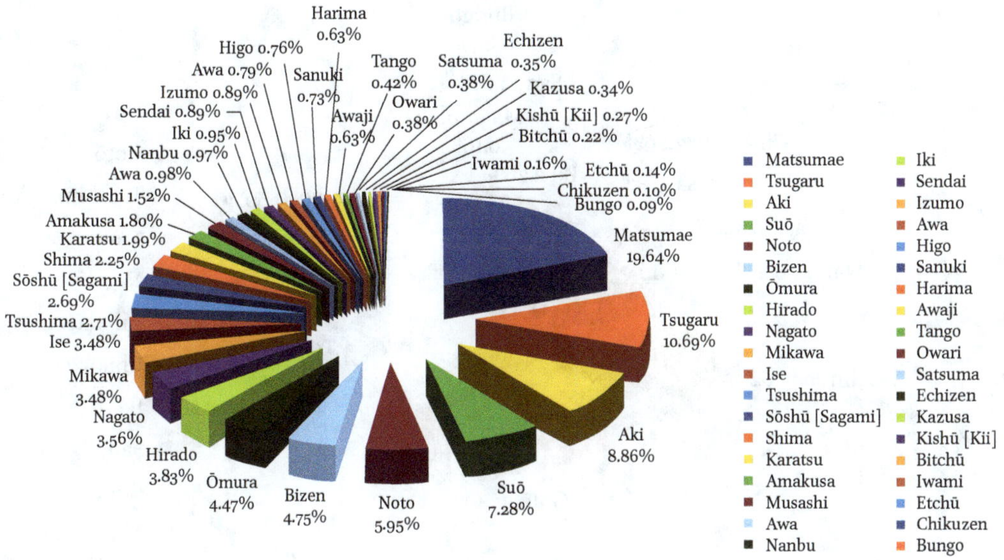

FIGURE 6.12 *Geographical origins of sea cucumber to Nagasaki in the early 19th century.*

TABLE 6.10 *Geographical origins of dried abalone* (*Units*: kin)[72]

Region	Place	Quantity
Present-day Hokkaidō region	Matsumae	210
Present-day Tōhoku & Hokuriku regions	Nanbu	9,100
	Sado 佐渡	2,230
Present-day Chūgoku region	Nagato	5,950
Present-day Kyūshū region	Chikuzen	3,050
	Bungo	860
	Amakusa	4,050
	Tsushima	20,770
	Iki	3,000
	Hirado	55,300
	Gotō五島	28,450

72 Ibid., 389.

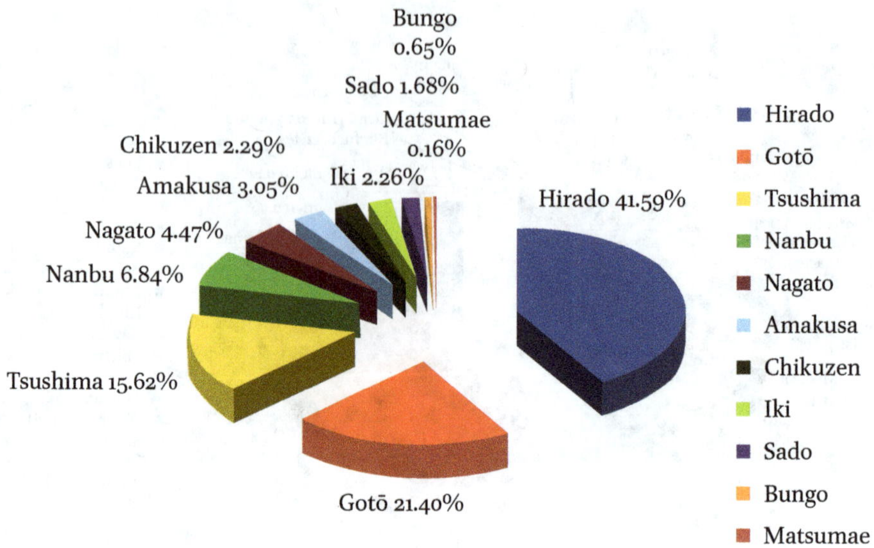

FIGURE 6.13 *Geographical origins of dried abalone.*

8 Japanese Dried Marine Products and Chinese Markets

While we know that goods were exported from Nagasaki to Chinese markets by Chinese merchant ships, what might the demand have been for Japanese goods in China? Though I have mentioned this point before, I would like to discuss it briefly here.[73]

Dried sea cucumber, made by extracting, boiling and drying the intestinal tract of sea cucumbers, was called "ocean ginseng" in Chinese, suggesting that it was as highly valued as ginseng (*ninjin* 人参) for its beneficial effects. Dried abalone had been regarded as an expensive gift since the Ming Dynasty. Dried shark fin was also widely regarded as a food item during the Ming Dynasty. Although poet and epicure Yuan Mei's 袁枚 *Suiyuanshidan* 随園食單 (printed in 1792) did not include seafood amongst its list of foods highly regarded since ancient times, it did contain a "sea food menu" that listed items that "now are popular," that is, highly regarded by Yuan's contemporaries. That list included items such as swallow's nest (*yanwo* 燕窩), sea cucumber, shark fin and dried abalone. These items were certainly dried marine products exported from Japan to China by Chinese merchant ships.[74]

73 See Matsuura, *Shindai kaigai bōekishi no kenkyū*, 388–97.

74 Ibid., 398.

While it is difficult to concretely show the reception of these goods in Qing period Chinese markets, the situation at the end of the dynasty can easily be seen in Meiji period Japanese consular reports such as the "Status of Marine Goods in Tianjin Market, China" (Shinkoku Tenshin shijō kaisanbutsu keikyō 清國天津市場海産物景況), a report from the Japanese imperial consulate in Tianjin dated 1886/8/28.[75]

According to this report, "eight or nine out of ten marine goods [imported to Tianjin] are produced in Japan. Although Korean sea cucumber and Taiwanese shark fin try to compete with Japanese products, at this time they generally are shut out of the market by Japanese products." Thus we see how Japanese dried sea cucumber overwhelmed that produced in Korea, and likewise Japanese shark fin vastly outnumbered that produced in Taiwan. In fact, most dried marine products originated in Japan.[76]

The same report describes the status of consumption of various marine products in Chinese markets (see table 6.11). Of the dried marine products exported from Japan in 1886, sea cucumber was highly regarded for its beautiful shape. Kelp became a daily comestible because of its supposed medicinal value, and it became popular even in Shanxi and Henan provinces. Shark fin with thick flesh was particularly well received. *Kanten* agar became popular in the Chinese interior, and *kanten* imported to Tianjin was often transported to Beijing, Shanxi, Shandong and Henan. Dried cuttlefish from Japan was popular, and that imported to Tianjin was transported to Bejing and Shanxi Province.

---

75 "Shinkoku Tenshin shijō kaisanbutsu keikyō," 46–49.

76 海産物ハ十中八九ハ本邦産ニシテ、就中朝鮮産煎海鼠、臺灣産鱶鰭ノ如キ従前ハ本邦産ニ對シ競争セシモ、目下殆ント消滅ノ色ヲ現ハセリ」とあるように、日本産の干し海鼠が朝鮮産を圧倒し同様に日本産の鱶鰭も台湾産の鱶鰭を凌駕していた。さらに昆布に関しては「日本ヨリノ輸入ト掲ケタルハ昆布一項ノミニシテ、其他ノ海産物ニ至リテハ支那内地産及外國産ノ輸入若干トノミ記載アリテ、特ニ本邦ヨリノ輸入年額ヲ記載セサルヲ以テ、本邦ヨリハ絶テ輸入ナキニ似タレトモ實際市上ニ陳列賣買スル海産物ハ多ク本邦産ノモノニシテ支那商人モ亦日本産ト稱呼スルヲ見レハ、特ニ海関報告中ニ外國産ト記載スルモノハ【其中多少魯國産ト南洋産アルニモセヨ】概シテ本邦産ニ係ルモノトスルモ、蓋シ大過ナカルベシ. [Kelp is the only product listed as an import from Japan, and only a few other marine products are recorded as domestic or foreign import, with none listed as coming from Japan; it may seem that there were absolutely no imports from Japan. Yet, in actual fact the large proportion of marine products laid out and traded on the market are produced in Japan, and the Chinese merchants even called them products of Japan. Accordingly, we may particularly imagine that those goods recorded as 'foreign' in the marine goods reports (notwithstanding those few listed as Russian or from the southern Chinese coastal islands) are largely Japanese; this would certainly not be a misstatement.] ibid., 46–49.

A considerable amount of dried abalone came from Taiwan, but the thick flesh and high-quality drying of the Japanese product meant that Japanese abalone found its way to Beijing, Henan and Shandong provinces. Although domestic Chinese varieties of seaweed and *tosaka* 鶏冠 seaweed were also brought to market in Tianjin, Japanese versions were regarded more highly and, like the above dried marine products, were shipped to Henan and other provinces. In particular, *tosaka* seaweed was transported even to coastal Shandong.[77]

Although these accounts date from the Meiji period, we may imagine that similar tastes drove the consumption of goods exported during the Edo period.

TABLE 6.11 *Status of consumption of various marine products in Chinese markets*[78]

Sea cucumber	Dried sea cucumber imported from Japan is predominantly of two varieties, number nine and number ten. Number ten is further divided into large and small. They are consumed in the greatest amount. Number nine is large, and perhaps because of its high price little is consumed. Those that have lots of spines and expand when boiled are highly regarded. At present, for every one hundred *kin* traded, number ten (large) fetches thirty-five *ryō*; number ten (small), twenty-six *ryō*; and number nine, forty-eight *ryō*. The so-called 'Koryŏ sea cucumber' (高麗参, Chinese *gongbieda* 工別大) is produced in Korea, and at present, one hundred *kin* of the large variety is worth twenty-three *ryō*, while the small is worth fifteen *ryō*. Sea cucumbers produced in Japan are highly regarded for their pleasing shape, while those from Korea are widely consumed for their low price and good flavour. The yearly value of these foreign imports is 55,430 *ryō* (The 1885 Maritime Customs Report, produced by Chinese officials, records the same value).

77 Ibid., 47–49.

78 Ibid., 47–49.

Kelp (Chinese *haidai* 海岱)	A deep green colour and complete freshness are valued. If [the kelp] is even slightly aged, consumption falls drastically. Around August and September exports to the Shanxi and Henan regions are greatest. Moreover, the locals of those regions believe kelp to help reduce phlegm, and not a few incorporate [kelp] into their daily diet. Before kelp from Japan began to be imported, kelp produced in the Guangdong region was imported. Since Japanese kelp began to be imported, that from Guangdong has been almost completely shut out of the market, perhaps because of the low price and good taste [of the Japanese product]. At present, very little Guangdong kelp is imported. At present, the value of kelp imports for every 100 *kin* is 2 *ryō* 4 *sen* 銭. The yearly value of imports is 121,301 *ryō*. Red kelp comprises 1,993 *ryō*.
Shredded kelp (Chinese *daisi* 岱絲)	Although there are two types of shredded kelp, rough and fine, the fine is best suited to Chinese tastes. At present, 100 *kin* is worth 1 *ryō* 8 *sen*, for a yearly value of 3,049 *ryō*.
Dried scallops (伊多剌貝; Chinese *ganbei* 乾貝)	Of all the marine products produced in Japan, the most highly regarded are dried scallops, in particular for their round shape and large size. Demand for this product generally comes from the Shandong and Shanxi regions. There are truthfully very few dried scallops produced in China, and no [Chinese-produced scallops] are imported here. When dried scallops were first imported, they were highly regarded and valued, at five hundred *ryō* per one hundred *kin*; however, more recently the prices have fallen considerably, perhaps due to oversupply. Current prices are sixteen *ryō* per one hundred *kin*.
Shark fin	Consumption of both black and white varieties is greatest. Regardless of the variety, large fins with thick flesh are highly sought after. However, the Chinese are displeased when including useless flesh at the base of the fin needlessly increases the product's weight.

TABLE 6.11 *Status of consumption of various marine* (cont.)

	Taiwanese shark fin has an excellent taste, and expands after boiling to become like snow. (It costs 110 *ryō* per 100 *kin*.) Despite high regard for it, its high cost keeps demand low. At present, 60 *ryō* of white shark fin to 40 *ryō* of black shark fin are sold (per 100 *kin* of shark fin, counted in *fuku* [副]; one *fuku* refers to the pair of fins behind the gills [頬]. The dorsal fin is not counted). There is a variety called *duichi* 堆翅, imported from the south, which is dried out after being cut. Consumption of this variety is very low, with no steady quantity being traded. The yearly value of foreign import runs to 26,341 *ryō* of white shark fin and 22,630 *ryō* of black shark fin; the value of imports from other Chinese ports runs to 6,464 *ryō* of white shark fin and, 2,946 *ryō* of black shark fin.
Kanten (寒天 agar) (Chinese *yangfen* 洋粉)	All the *kanten* imported from Japan is shredded *kanten* 絲寒天. Large, square items arrived as samples, but prove difficult to sell. For the most part consumption is very low in Tianjin, and the bulk is exported to the Beijing, Shanxi, Shandong and Henan regions. There are three types of *kanten*: top grade 頭番, second grade 二番, and third grade 三番. White colour is also valued. Of the three, top quality agar is pure white, and thus consumed in greatest quantity. At present, top quality agar fetches 17 *ryō* 5 *sen* per 100 *kin*, and the yearly value of foreign imports is 19,515 *ryō*.
Dried cuttlefish (Chinese *youyu* 魷魚)	Large size and thick flesh are valued. In previous years, cuttlefish from Taiwan were in high demand compared to those from Japan, due to their large size (valued at 13 *ryō* per 100 *kin*). Yet they have been overwhelmed by Japanese imports, and recently only Japanese products are traded. A large proportion is exported to Beijing and Shanxi. At present, 100 *kin* fetches 8 *ryō*, and the yearly value of foreign imports is 2,173 *ryō*.

Abalone (Chinese *baoyu* 鮑魚)	Amongst the imports from Japan, there are three varieties: large, medium and small, of which the medium size sells in greatest quantity. A dried exterior, thick flesh and yellow-black colour are highly valued. A large quantity is exported to the Beijing, Shanxi and Henan regions. In previous years Taiwanese goods were imported, but were overwhelmed by Japanese products, and thus did not sell well. At present, one hundred *kin* fetches twenty-three *ryō*.
Seaweed (Chinese *zicai* 紫菜)	A large size with red bands, and unadulterated by sand, is highly prized. Most is exported to Beijing and Henan, with little consumed here [in Tianjin]. Because [seaweed] is transported from Ningbo, some does arrive here by ship. Nevertheless, it is only a small amount, which does not play a significant role in the market. At present, 100 *kin* garners 20 *ryō*, with a yearly value of 20,434 *ryō*.
Tosaka 鶏冠 seaweed	[Seaweed] with a red band and unadulterated with sand is highly regarded. Much is exported to Henan and Shandong provinces. At present, every one hundred *kin* sold at market receives between three and five *ryō*. The so-called '*lincai* 鄰菜', with its round shape, reminds one of *hongcai*. Consumption is high of products from Guangdong and Xiamen. Yet it does not approach that of our *tosaka* seaweed. At present, one hundred *kin* barely fetches between three and six *ryō*.

9 Concluding Remarks

Historical export records reveal the changes in export volumes of marine products, particularly sea cucumber, dried abalone and shark fin, from Nagasaki to China across the Edo period. The *Tōsen*, as Chinese merchant ships were called in Japan, often had two or three enormous masts to carry out their seafaring trade, and required stabilizing ballast for long-distance transport. For ships travelling to Nagasaki from Qing China, large quantities of sugar served as ballast.[79] In addition, the Chinese ships transported silk fabrics, books, and Chinese medicine. Chinese vessels returned home with cargoes of Japanese copper, eagerly sought after by the Qing government as the principal ingredient with which to mint coins. Known as "foreign copper" in China, Japanese copper constituted a significant and important proportion of China's copper supply sources during the early Edo period. Yet as time passed and Japanese copper production steadily declined, the amount of copper in the return cargoes of Chinese ships leaving Nagasaki for China decreased, until by the mid- and late-Edo periods the Japanese authorities limited copper exports to one hundred thousand *kin* per ship.

In contrast, increasing demand in China for dried marine products made them a significant new export. Dried sea cucumber, dried abalone and shark fin bear particular attention as the three principle dried goods, representing a significant proportion of exports from the mid- to late Edo period. Above all the three dried goods of dried sea cucumber, dried abalone and shark fin were essential ingredients for the seafood cuisine that had grown popular from the mid-Qing period onwards. In addition, kelp and dried cuttlefish, as well as various sea plants such as *tengusa* seaweed, played an important role.[80] From the perspective of the Chinese junk trade, we may say that the decline in copper exports was balanced out by an increase in dried marine products.

---

79 Matsuura, "Edo jidai Tōsen ni yoru satō yu'nyū to kokunai shōhi no tenkai," 335–59.

80 Matsuura, "Nisshin bōeki ni yoru tawaramono no Chūgoku ryūnyū ni tsuite," 19–38, and Matsuura, *Shindai kaigai bōeki shi no kenkyū*, 382–402.

CHAPTER 7

The Import of Chinese Sugar in the Nagasaki Junk Trade and Its Impact

Matsuura Akira 松浦章

1 Introduction

An exemplary sweetener consumed around the globe, sugar seasons daily cooking and sweetens drinks in many regions. Yet it was not until production reached surplus levels around the world after the First World War that sugar entered the daily lives of people around the world.[1]

During the Edo period Chinese junks, called *Tōsen* 唐船, carried large amounts of Chinese sugar to Japan every year. A large part of this sugar was produced in the coastal regions of eastern Guangdong 廣東 and southern Fujian 福建 provinces, and in Taiwan. In the early Edo period ships transported sugar directly from these production areas to Japan; in the mid- and late Edo periods, coastal merchant ships transported sugar from these areas first to Zhapu 乍浦 in Zhejiang Province 浙江省, where it was transferred onto junks bound for Nagasaki. Much of the sugar shipped to Nagasaki was transported onwards to Osaka and then to the rest of Japan.

In the early 18th century, sugar cane harvesting began in Japan. The eighth Tokugawa shogun, Yoshimune (吉宗, r. 1716–1745), famously created the impetus to spread techniques for harvesting sugar cane, the basic ingredient of sugar. As a result, sugar consumption expanded during the Kyōhō years.[2] According to the record of Yoshimune's activities, the *Yūtokuin gojikki furoku* 有徳院御實記附録 (volume 17), Yoshimune explored ways to reduce Japan's dependence on imports from China for the bulk of domestically consumed sugar.[3] Yoshimune considered possibilities for domestic sugar production, particularly methods for harvesting sugar cane. In 1727 (Kyōhō 12), he approached

* Paper translated by Michael Thornton.

1 Sekino, *Sekai tōgyō bunka shi*, 3.

2 Kobata, "Satō no shiteki kenkyū ni tsuite," 217. This paper is a revised version of an essay with the same title in *Taiwan jihō* 臺灣時報 186 and 187 (1935/5, 6). It is one of the most important fruits of historical research on the import and spread of sugar in Japan.

3 *Tokugawa jikki* 9, 316.

 | DOI 10.1163/9789004304512_008

the head of Shimazu domain 島津藩, Matsudaira Tsugutoyo 松平繼豊 (1701–1760) to study harvesting techniques in Satsuma. Furthermore, he transplanted sugar cane to Nagasaki and Suruga 駿河. He also studied Chinese texts on sugar cane harvesting technologies, ordering the magistrate in charge of imported books, Fukami Arichika 深見有隣 (1691–1773), to conduct a survey of relevant Chinese literature. Yoshimune also sought to learn about sugar production from merchants who arrived in Nagasaki. In other words, records show that Yoshimune used three methods to study sugar cane harvesting: observations of the cane harvest in Ryūkyū and Satsuma; study of Chinese texts; and conversations with Chinese merchants in Nagasaki. In this way, not only did overall sugar production increase, but the quality of sugar also improved.

In this essay I shall discuss the import of sugar from China during the Edo period, and also the increase in sugar consumption in Japan. Furthermore, I will investigate the reception of Chinese culture through the lens of Chinese sugar imports during the Edo period, and in turn the Japanese transformation of that culture.

2 Chinese Sugar Imports in the Records of the Dutch Trading House at Nagasaki

Chinese records do not detail the quantity of sugar imported to Nagasaki, so I turn to the records kept by the Chinese merchants' Dutch competitors.[4] Dutch traders were keenly aware of the Chinese ships entering Nagasaki, and kept meticulous daily records of the ships and their cargo capacity. Here I look at only those records that deal with sugar.

An entry dated 1641/7/5 in the *Nederlandsche Oost-Indische Compagnie, Comptoir Nangasackij* (長崎オランダ商館の日記 [Diary of the Dutch trading house at Nagasaki], henceforth *Diary*) states that merchant Zheng Zhilong 鄭芝龍 (1604–1661) sent twelve ships laden with sugar to Nagasaki in 1641.[5]

On 1641/7/10, an entry records that "early this morning, one small junk arrived in port from Fuzhou 福州. Its cargo contained 16,000 *kin* of muscovado (*kurozatō* 黒砂糖, a type of unrefined brown sugar)... and 400 *kin* of white

4 Details of sugar imported by Chinese and Dutch ships during the Edo period may be found in Iwao, "Edo jidai no satō bōeki ni tsuite," 1–33. Iwao exhaustively explores the Dutch records.

5 "According to one merchant, this year 'the official' is sending twelve ships of sugar here [Nagasaki]. The first entered the port around noon. Its cargo... included 19,800 *kin* of sugar...." Here 'the official' refers to Zheng Zhilong. 1 *kin* 斤 is approximately 600 grams". See *Nagasaki Oranda shōkan no nikki* 1, 56.

sugar (*shirozatō* 白砂糖)."[6] The same entry later mentions that "late at night a Chinese vessel arrived from Canton (Guangdong). Its cargo included 11,500 *kin* of white sugar, and 1,000 *kin* of muscovado."[7] On 7/12, "in the afternoon, the official's second ship arrived with 270,000 *kin* of white sugar."[8] On 7/13, "in the morning, a junk from Fuzhou arrived, laden with 70,000 *kin* of white sugar."[9] On 7/14, "two of the official's junks from the aforementioned location [Fuzhou] entered port, with a cargo of…139,200 *kin* of white sugar, 10,300 *kin* of muscovado, [and] 30,000 *kin* of rock candy (*koorizatō* 氷砂糖)."[10] On 7/22, "in the evening, a Chinese junk from Fuzhou arrived. Its cargo included 4,200 *kin* of white sugar."[11] On 7/23, "shortly after noon, three junks entered port. The cargo of one ship from Guangnan 廣南 was 40,400 *kin* of muscovado."[12] As for the others, "the two ships from Fuzhou carried 77,050 *kin* of white sugar, [and] 8,300 *kin* of muscovado."[13] According to an entry from 7/24, "another of the official's junks from Zhangzhou 漳州 arrived. Its cargo included 1,660 *kin* of muscovado."[14] On 7/25, "at noon, four small junks arrived carrying the following goods: 275,700 *kin* of white sugar, 4,800 *kin* of muscovado, and 62,300 *kin* of rock candy."[15] On 7/26, "in the morning and afternoon five junks arrived in port. The three ships from Canton carried 55,000 *kin* of white sugar, and 1,200 *kin* of rock candy."[16] The remaining "two ships from Tonkin carried the following goods,"[17] though the records do not list sugar cargoes. According to an entry for 7/27, "today two junks arrived. One came from Canton, carrying 2,500 *kin* of white sugar,"[18] while "the ship from Quanzhou 泉州 carried 18,500 *kin* of white sugar."[19] On 7/29, "a large junk arrived from Quanzhou. Its cargo included 10,000 *kin* of white sugar and 18,000 *kin* of muscovado."[20]

---

6 Ibid., 57.
7 Ibid., 58.
8 Ibid., 58.
9 Ibid., 58.
10 Ibid., 58.
11 Ibid., 62.
12 Ibid., 63.
13 Ibid., 64.
14 Ibid., 65.
15 Ibid., 66–67.
16 Ibid., 67.
17 Ibid., 67.
18 Ibid., 68–69.
19 Ibid., 69.
20 Ibid., 71.

Chinese junks were not the only vessels shipping sugar to Nagasaki; the Dutch also carried sugar cargoes. On 7/30, "around nine in the morning, we started unloading the orange trees. By evening the textiles had mostly been finished, leaving only forty barrels of sugar [to be unloaded]."[21]

Records of the type and amount of goods brought to Nagasaki by Chinese vessels lie in the entry for 1641/10/11. On eighty-nine junks from China arrived "sugar: 5,427,500 *kin*; muscovado: 251,700 *kin*; rock candy: 47,300 *kin*."[22] Moreover, "three junks from Guangnan carried 4,000 *kin* of muscovado, [and] 20,000 *kin* of white."[23] Of the above totals, "the official wrote a memorandum of goods imported on six large junks," recording "4,000 *kin* of sugar and 35,000 *kin* of muscovado."[24]

The *Diary* entry for 1642/10/16 states that thirty-four merchant ships from China imported 32,800 *kin* of rock candy, 24,000 *kin* of [white] sugar, and 160,100 *kin* of muscovado.[25] According to the entry for 1644/11/15, 54 Chinese merchant vessels carried 489,800 *kin* of white sugar, 849,600 *kin* of muscovado, and 78,150 *kin* of rock candy to Nagasaki.[26] The following year, based on an entry dated 1645/11/25, seventy-six Chinese vessels brought 54,800 *kin* of rock candy, 1,770,000 *kin* of white sugar and 1,553,000 *kin* of muscovado to Nagasaki.[27] According to an entry for 1646/10/27, fifty-four ships transported 779,500 *kin* of white sugar, 145,500 *kin* of rock candy, and 258,100 *kin* of muscovado to Nagasaki over the course of the year.[28] "Between the end of the seventh month and 9/20," according to an entry for 1648/12/8, seventeen Chinese merchant ships transported 12,000 *kin* of white sugar, 91,000 *kin* of muscovado and 83 *kin* of rock candy to Nagasaki's market.[29] On 1649/11/5, the *Diary* records a total of fifty merchant ships carrying 685,800 *kin* of muscovado and 51,450 *kin* of powdered sugar (*konazatō* 粉砂糖);[30] according to an entry for 1650/10/25, some seventy ships imported 790,960 *kin* of white sugar and muscovado, as well as 6,150 *kin* of rock candy that year.[31] Between 1652/11/10 and 1653/11/10, fifty-four Chinese merchant ships carried 152,100 *kin* of Chinese sugar, 584,870

21 Ibid., 71.
22 Ibid., 108.
23 Ibid., 110.
24 Ibid., 111.
25 Ibid., 194, 197.
26 Ibid., 372, 374–75.
27 *Nagasaki Oranda shōkan no nikki* 2, 62–63.
28 Ibid., 103, 106.
29 Ibid., 224, 226–27.
30 Ibid., 265–66.
31 Ibid., 321–22.

TABLE 7.1 *Volume of sugar cargo on Chinese ships serving Nagasaki, 1642–1653* (*in* kin)

Year	Total ships	Sugar 砂糖	White sugar 白砂糖	Muscovado 黒砂糖	Rock candy 氷砂糖	Total	Average load
1642	34	24,000		160,100	32,800	216,900	6,379.4
1644	54		489,800	849,600	78,150	1,417,550	26,250.9
1645	76		1,770,000	1,553,000	54,800	3,377,800	44,444.7
1646	54		779,500	258,100	145,500	1,183,100	21,909.3
1648	17		12,000	91,000	83	103,083	6,063.7
1649	50		51,450	685,800		737,250	14,745.0
1650	70	790,960			6,150	797,110	11,387.3
1653	54	152,100		584,870	37,250	774,220	14,337.4
Total	409	967,060	3,102,750	4,182,470	354,733	8,607,013	
Average	51.1	6,120.6	12,362.6	12,338.7	988.1		21,044.0

kin of muscovado, and 37,250 *kin* of rock candy to Nagasaki market, according to an entry from 1653/11/12.[32]

As seen above, Chinese merchants imported large amounts of sugar to Nagasaki every year. I would now like to consider how much each of these ships would have taken on as cargo.

As seen in table 7.1, between 1642 and 1653 yearly sugar imports on Chinese merchant vessels reached a maximum amount of 3,370,000 *kin* (ca. 2,022 tons), with cargoes in the leanest years still exceeding 700,000 *kin* (ca. 420 tons). The sugar represented by these values was all consumed domestically within Japan.

From the perspective of Chinese merchant ships travelling to Japan, even a lean year saw each ship carry roughly 6,000 *kin* of various types of sugar. Averaging these eight years, each ship carried roughly 12,000 *kin* of white sugar, 12,000 *kin* of muscovado and roughly 1,000 *kin* of rock candy, for a total of approximately 25,000 *kin* of sugar per cargo load. Given that one *kin* approximates 600 g today, each cargo works out to 15 tons. This was clearly the most appropriate ballast weight for Chinese sailing ships. Moreover, this demonstrates that demand for such a quantity of sugar existed in Edo-period Japanese society.

32 *Nagasaki Oranda shōkan no nikki* 3, 248, 250.

3 Japanese Records of Nagasaki Sugar Imports

As described in the sixth volume of *Laoxuean biji* 老學庵筆記 by Sung-era poet Lu You 陸游 (1125–1209), sugar was imported to China by foreign envoys during the reign of the Taizong emperor (太宗, r. 626–649) of the Tang Dynasty. Thereafter sugar spread widely throughout China.[33] By the Edo period, Chinese sugar was regularly imported to Nagasaki on trading ships.

3.1 *Japanese Records of the Import of Chinese Sugar*

I would now like to investigate how the Japanese records portray the sugar cargoes of Chinese merchant vessels, in contrast to the records of the Dutch discussed above.

In the seventh month of 1686, a cargo ship from Xiamen 廈門 drifted ashore on Tsushima 対馬. Captain Chen Ang 陳昂 recorded a cargo of "4,000 *tan*[34] of white sugar and 2,000 *tan* of rock candy."[35]

We may summarize as follows the records of sugar extracted from the accounts of cargo imported to Nagasaki in 1729. These accounts are listed in the *Getsudō kenbunshū* 月堂見聞集, which recorded the comings-and-goings and hearsay of the Genroku 元禄 (1688–1704) through late Kyōhō periods, focusing on Edo, Kyoto and Osaka but covering the whole country.

Particularly noteworthy is the 67,000 *kin* of sugar transported on the No 17 ship from Tonkin (today Hanoi), which amounts to forty tons.

The cargo of a vessel shipwrecked upon Hachijō-jima 八丈島 on 1753/12/10, owned by Gao Shanhui 高山輝, included sugar. Despite losing a portion of the cargo to rough weather, "5,040 *kan*[36] of rock candy and 8,901 *kan* of white sugar" remained.[37] This ship's cargo was transferred to two Japanese ships, the *Wagōmaru* 和合丸 and *Ōsugimaru* 大杉丸, whose cargo records tell us that the vessel shipwrecked on Hachijōjima carried both white sugar and rock candy.[38]

---

33 *Laoxuean biji*, 80.

34 1 *tan* [擔, 担] = 100 *kin*.

35 *Ka-I hentai*, jō, 643; *Tsūko ichiran* 5, 325.

36 1 *kan* [貫] is 3.75 kilograms.

37 "Gokaen zuihitsu", 56.

38 "Catalogue of Cargo for the Wagō-maru and Ōsugi–maru
- White sugar: 163 *ken* 件 (and 8 *tsutsumi* 包 for food supplies)
- Rock candy: 132 *ken* (and 1 *tsutsumi* for supplies)

Catalogue of Remaining Goods
- White sugar: 306 *ken*
- Rock candy: 44 *ken*"

TABLE 7.2 *Cargo carried on ships entering Nagasaki harbor, tenth month*

Ship number	Origin	Date of entry	White sugar (in *kin*)	Muscovado (in *kin*)
8	Nanjing 南京		330	250[39]
9	Ningbo 寧波	10/22	15,000[40]	
10	Ningbo	10/23	18,000[41]	
11	Nanjing		18,000[42]	
15	Ningbo		22,500[43]	
16	Champa 占城		23,200[44]	
17	Tonkin 東京		67,000[45]	

On 1780/4/30, a trading ship owned by Shen Jingzhan 沈敬瞻 drifted ashore near Chikura 千倉, in Awa Province 房州. Combining the 262,500 *kin* of white sugar and 12,500 *kin* of rock candy, we reach a total of 275,000 *kin*—or 165 tons.[46] That a single ship imported this much sugar into Japan is noteworthy.

On 1807/1/7, a trading ship owned by Wang Zongding 王宗鼎 and bound for Nagasaki drifted into Chōshi Bay 銚子浦 in Shimōsa 下総. Its cargo was reported in "A general and detailed account of the cargo of the *Jinyuansheng* 金源盛." Adding together the highest-quality Chinese *sanbon* sugar 三盆 and

Ken is an unknown measure, but probably refers to a quantity of goods placed into large sacks; the weight of those goods is also unclear. *Tsutsumi* were smaller bags, and likely contained some fraction of a *ken*. See "Junkairoku," 69–70.

39 *Kinsei fūzoku kenbunshū* 2, 174; *Tsūkō ichiran* 6, 36.

40 *Kinsei fūzoku kenbunshū* 2, 175; *Tsūkō ichiran* 6, 8.

41 *Kinsei fūzoku kenbunshū* 2, 175; *Tsūkō ichiran* 6, 9.

42 *Kinsei fūzoku kenbunshū* 2, 175; *Tsūkō ichiran* 6, 36.

43 *Kinsei fūzoku kenbunshū* 2, 175; *Tsūkō ichiran* 6, 9.

44 *Kinsei fūzoku kenbunshū* 2, 176.

45 Ibid., 176.

46 "A Nanjing trading vessel drifted into the waters of Awa Province. The magistrate at Uraga 浦賀 and shogunal officials investigated the cargo of the ship, taking an aide from the *Rokujukken yakushu ton'ya* trading house in Edo, and submitted this report.
Memorandum:
- White sugar: 262,500 *kin*
- Rock candy: 50 *oke* (桶, 12,500 *kin*)

The latter was taken for fuel and burnt."
See "Zokudankai," 94.

sentō sugar 泉糖 from Quanzhou gives a total of 67,857 *kin*, or more than forty tons.[47]

In 1809 (Bunka 6, Jiaqing 14), the seventh ship of the year carried 18,300 *kin* (200 bags (*tsutsumi* 包)) of rock candy, 32,000 *kin* (265 bags) of *Sanbon*, and 45,550 *kin* (398 bags) of *sentō* sugar. This adds up to a total of 95,850 *kin*, or more than fifty-seven tons of sugar.[48]

In 1826/1, the ship *Detai* 得泰 drifted into Suruga, whence it was towed to Nagasaki and numbered the eighth ship of 1825. Owned by Ningbo merchants Liu Jingyun 劉景筠 and Yang Qitang 楊啓堂, the ship carried a cargo record (*tsūsen kasū* 通船貨数) that listed various types of sugar amongst its "670 items of cargo"[49]:

– white cake:	rock candy, 100 bundles [連 *ren*]
– *chōban* 頂番 sugar:	superior grade white sugar, 300 bags [包]
– *sanbon*:	high-grade sugar, processed as *sanbon*-grade, 650 bags
– *sentō* 泉糖 sugar:	middle grade; "Quan" refers to the place of production Quanzhou 泉州, 500 bags

The ship carried rock candy, *sanbon*, superior grade '*chōban* sugar,' and sugar from Quanzhou. There are no weights given for these amounts, but if we use the conversions from the seventh ship of 1809, we may calculate the values for the seventh ship of 1826: 200 bags of rock candy equaled 18,300 *kin*, meaning one bag equaled 91.5 *kin*; 260 bags of *sanbon* equaled 32,000 *kin*, meaning one bag equaled 123 *kin*; and 398 bags of *sentō* sugar came out to 45,550 *kin*, meaning one bag equaled 114 *kin*.

Thus for the *Detai*, 650 bags of *sanbon* works out to 79,950 *kin*; 500 bags of *sentō* sugar to 57,000 *kin*; if we presume *chōban* sugar weighed 100 *kin* per bag, then we reach a total of 30,000 *kin*; and similarly we may suppose that

---

47 "An investigation into various goods loaded onto the ship. The catalogue:
- 250 units of Number One high-grade white sugar [*sanbon* 三盆], weighing 28,224 *kin* [16.9 tons].
- 350 units of Number Two Quanzhou sugar, altogether weighing 39,633 *kin* [23.8 tons]."

See "Bunka teibō Tōsen hyōchakuki", 6.

48 See *Matsuura, Shindai kaigai bōekishi no kenkyū*, 377.

49 Tanaka and Matsuura, *Bunsei kyūnen Enshū hyōchaku Detai sen shiryō*, 28–30.

100 bundles of rock candy, at roughly 100 *kin* a bag, totaled 10,000 *kin*. This adds up to 176,950 *kin*, or slightly more than 106 tons. This ship's cargo contained more than 100 tons of various types of sugar.

Turning to the Dutch records, an investigation into the quantity of sugar transported to Nagasaki by Chinese ships in the late Edo period (Tenpō 天保 2–3; 1831–1832) yields the following results:

TABLE 7.3 *Amount of sugar imported by Chinese vessels in 1831–1832 (Tenpō 2–3)*[50] (*in* kin)

[Ship number]	Rock candy	White sugar	Superior-grade sugar (最上砂糖)	Standard-grade sugar (並砂糖)	Totals
1831 (Tenpō 2)					
1	36,000	124,700			160,700
2	29,500	112,385			141,885
3	27,696				27,696
5	34,700	197,500			232,200
6	20,000	130,000			150,000
7	20,000	163,000			183,000
8	15,000	134,000			149,000
9	24,200	180,000			204,200
10	25,000	190,000			215,000
1832 (Tenpō 3)					
5	16,000	126,000			142,000
6	11,500		50,000	35,000	96,500
7	10,000		62,000	45,000	117,000
8	40,000		39,100	40,000	119,100
9	11,477		67,500	40,000	118,977
Totals	321,073	1,357,585	218,600	160,000	2,057,258
Average (per ship)	22,933.8	150,842.8	54,650	40,000	156,120.1

Note: The average per ship excludes the third ship of Tenpō 2.

50 Nagazumi, *Tōsen yushutsunyūhin sūryō ichiran 1637–1833 nen*, 249–52.

The Chinese merchant ships travelling to Nagasaki in 1831 and 1832 carried 156,000 *kin* of sugar, white sugar and rock candy to Japan, nearly 100 tons.

3.2 *The Collection of Chinese Sugar Products and the Structure of Exports to Japan*

Now I turn to the question of how these varieties of sugar were collected and transported to Japan. For the Chinese merchant vessels that plied Nagasaki, the port of Zhapu in Zhejiang Province served as the centre of the Japan trade for roughly one hundred years between the mid-eighteenth century and the last years of the Tokugawa Shogunate.[51] This relationship between Zhapu and Japan in the Qing era was clearly recorded in the Daoguang-period *Zhapu beizhi* 乍浦備志, volume 14, "Qianming wobian 前明倭變," edited by Zou Jing 鄒璟. According to this record, Japanese copper was essential for China's domestic minting. In order to purchase this copper, state merchants (*guanshang* 官商) were established at Zhapu, from where they headed eastwards for Japan.[52] Furthermore, the same document records details of the trade structure: a state and civil division of merchants was created, and each was allowed three ships a year. In midsummer, between *geshi* 夏至 and *shōsho* 小暑, a total of six vessels transported sugar and miscellaneous goods desired by the Japanese from Fujian and Guangdong east to Japan.[53] The period of trade with Japan, *geshi* to *shōsho*, corresponds roughly to 20 June through early July in today's calendar. During these twenty days, ships left Zhapu for Japan. The document further states that if westerly winds blew regularly, the voyage lasted four to five days.[54] If the winds were not favourable, the voyage required at least ten and up to thirty or forty days. In general, these ships returned in the middle of the ninth month carrying Japanese copper and dried marine products such as kelp and dried sea cucumber.[55]

The ships then made a second voyage. Once they had unloaded the goods brought back from Japan, the vessels loaded up with sugar and other goods and once more set sail for Japan in the twenty days or so between *shōsetsu* 小雪 and *taisetsu* 大雪—between the end of November and early December in today's terms. They then returned to Zhapu in the fourth or fifth month of the following year.[56] They carried Japanese copper and other miscellaneous

51 Matsuura, *Shindai kaigai bōekishi no kenkyū*, 98–117.

52 *Zhongguo difangzhi jicheng: xiangzhen zhi zhuanji*, 229.

53 Ibid., 229–30.

54 Ibid., 230.

55 Ibid., 230.

56 Ibid., 230.

goods on this return voyage as well. In this way they set a semi-annual pattern of trading voyages. Moreover, the documents state that every year 1,200,000 *kin* of Japanese copper was shipped to China, with each ship carrying 100,000 *kin* (ca. thirty tons).

Details about the sugar collected in Zhapu can be found in the sections "Guanliang 關梁" and "Haiguanshuikou 海關稅口" in the *Zhapu beizhi* (volume six). Sugar products brought to Zhapu were exported to the Hangzhou 杭州, Jiaxing 嘉興, Huzhou 湖州 and Jiangnan 江南 regions. Notably, during the Qianlong era (1736–1795) two-thirds of the sugar transported to Zhapu originated in Guangdong. Most of the sugar merchants in Zhapu were said to be people from Chaozhou 潮州, suggesting that the bulk of sugar traded in Zhapu was produced in the Chaozhou region of Guangdong. These merchants stayed in Zhapu the entire year. As ships laden with sugar arrived, transport workers known as *guotanghang* 過塘行 performed the customs procedures. In later years, it became common for most sugar to be transported to Shanghai rather than Zhapu.[57]

The records of a Ryūkyū official charged with investigating the wreck on Yaeyamajima 八重山島 of a merchant vessel from Chaozhou, dated 1814/12/25, appear in the *Rekidai hōan* (歷代寶案, 2: 118). This 149th Chenghai ship, a merchant vessel owned by Wu Yongwan 呉永萬 of Chenghai 澄海 County in Chaozhou Prefecture, left Donglong 東隴 port on 6/18 with fifty-eight people onboard (twenty-two passengers and thirty-six sailors, including the captain), carrying a cargo of muscovado and white sugar. On 8/7 it arrived in Tianjin Prefecture 天津府, where it sold its cargo.[58] From this example we may conclude that ships from Chaozhou regularly plied the seas off northern China with sugar cargoes.

This pattern is also evident in the chapter "Customs and Occupations (Fengsu Shengye 風俗, 生業)" of the Jiaqing-era *Chenghai xianzhi*, which describes how sugar was gathered from every village in Chenghai County during the harvesting season, and loaded onto ships during the third and fourth months when southerly winds blew. The ships then travelled to Suzhou 蘇州 and Tianjin, where they reaped tremendous profits. Accordingly, we may imagine that much of the sugar from Chenghai County was exported to Zhapu.[59]

---

57 Ibid., 149.

58 Matsuura, "18–19 seiki ni okeru Nansei shotō hyōchaku chūgoku hansen yori mita shindai kōungyō no issokumen", 17–75.

59 Chenghai-xianzhi vol. 6: 8.

According to a palace memorial from Jiangnan provincial military commander Lin Junsheng (提督江南總兵官左都督 林君陞),[60] dated 1753/7/4, sugar ships from Fujian and Guangdong arrived at the ports of Liuhe 劉河, Chuansha 川沙, Wusong 呉淞, and Shanghai to trade during the season of southerly winds, in the fourth and fifth months. In the ninth and tenth months they purchased cotton and returned.[61] In this way, sugar produced in Chinese southeastern seaboard regions was transported to ports near Zhapu and Shanghai by ocean-going vessels from Guangdong and Fujian. Part of that sugar was transported from Zhapu to Japan by Nagasaki-bound trading ships.

4 The State of Sugar Consumption in Japan

Now we turn our focus to the consumption of sugar in Japan: how was the sugar brought to Japan by Chinese ships consumed?

4.1 *Sugar Imported to Japan*

Terashima Ryōan 寺島良安 wrote about sugar in a glossary appended to the passage on sugar cane (*kansho* 甘蔗) in the *Wakan sansai zue* 和漢三才図会 (volume 9). According to Terashima, at the beginning of the 18th century 2,500,000 *kin*, or ca. 1500 tons, of white sugar was imported annually. This sugar was transported not only by Chinese vessels, but also by Dutch ships. Sugar was roughly divided into three types: rock sugar, white sugar and muscovado. These were all produced from sugarcane, which at the time did not grow widely in Japan. For that reason, sugar was imported through Nagasaki. White sugar imports totaled 2,500,000 *kin*, with Taiwanese sugar regarded most highly, followed by sugar from Jiaozhi [Cochin] in southern Vietnam, then sugar from Nanjing, Fujian and Ningbo. Indonesian sugar, imported from the Netherlands, was called "Dejima White" for its association with the Dutch settlement at Dejima in Nagasaki, and regarded as the lowest quality.[62]

---

60 Lin Junsheng was concurrently Left Commissioner-in-chief (*zuodudu* 左都督) in the Board of War, and also regional commander (*zongbingguan* 總兵官) of Zhenhai 鎮海 and Dinghai 定海.

61 A palace memorial submitted by Lin Junsheng, dated 1753/7/4, in [GDQZ], 5, 689.

62 "In my view, there are three types of sugar: rock sugar (*kōrizatō*), white sugar (*shirozatō*) and muscovado (*kurozatō*). Originally these are all the same thing, and differ only in the way that pig iron (*zuku*), iron (*kurogane*) and steel (*hagane*) differ. In our country, even sugarcane that is transplanted does not grow well. As a seasoning for rice cakes (*mochidango*), sugar is considered essential. The general outline of imports from other countries is as follows:

Terashima described some foods that used sugar in "Candied sweets" (*satōzuke no kashi* 沙糖漬の菓子):

> In my view, items such as tangerines (*mikan* 蜜柑), citron (*busshukan* 佛手柑), dried asparagus root (*tenmondō* 天門冬), ginger (*shōga* 生薑) and winter melon (冬瓜 *tōgan*) are candied in sugar and made into sweets. Yet in order to preserve them for several months, they are soaked in limewater overnight and washed clean before being candied....[63]

Sugar thus served as an essential ingredient in important processed foods such as candied fruit.

As mentioned earlier, Tokugawa Yoshimune made great efforts to improve domestic sugar supply. In addition to ordering Fukami Arichika to comb imported texts for information on sugar, Yoshimune also had him gather information about sugar production from the captains of Chinese vessels arriving at Nagasaki. Li Daiheng 李大衡, of the sixth Xiamen ship, which arrived in the ninth month of 1726, recorded "Methods of Producing Muscovado 煮烏糖法" and "Methods of Producing White Sugar 煮白糖法." "Methods of Producing Muscovado" describes steps from the cutting of sugar cane and bamboo to the production of muscovado; "Methods of Producing White Sugar" describes several further steps including the addition of "ash made from barnacle shells" and further boiling to produce refined white sugar.[64]

Li Daiheng had also travelled to Japan on the twenty-third Xiamen ship in 1723. While the ship originated in Xiamen, it took on thirty-nine people in Shanghai, left that port on the 1723/11/23, and arrived at Nagasaki on 11/28. A record from that visit indicates Li Daiheng had been a secretary in charge of financial affairs on the twelfth ship of 1721. The trade permit (*shinpai* 信牌), which Li carried in 1723, had been issued to his fellow passenger, merchant

---

Approximately 2,500,000 *kin* (1 *ton* is 175 *kin*. At Nagasaki it is divided into two crates (*hako*), each containing 86.5 *kin* [*sic*]) of white sugar arrive from various foreign countries. That which is pure and not damp is good. It contains large clumps of round, flat sugar in the shape of rice cakes. This sugar is called *sanbon*. When broken apart, it becomes very white. Everything from Taiwan is considered the highest grade, followed by sugar from Jiaozhi. The next grade comprises sugar from Nanjing, Fujian and Ningbo, while sugar from Calapa [咬留巴 present-day Jakarta] and Holland (阿蘭陀 [referring to Dejima]) is considered of the lowest quality."

See *Wakan sansai zue* 16, 81.

63 *Wakan sansai zue* 18, 244.

64 *Tsūkō ichiran* 6, 24–27.

Yan Qizong 顏啓惣, in 1721.[65] We can confirm Li Daiheng's arrival in Japan on the following ships shown in table 7.4:

TABLE 7.4 *Ships on which Li Daiheng arrived in Japan*[66]

Year, ship number, origin	Ship owner	Notes
1721, Number 12, Xiamen	Zhou Yuanji 周元吉	Permit holder: Wu Chuyu 呉楚譽 Secretary: Li Daiheng
1723, Number 23, Xiamen	Li Daiheng	Permit holder: Yan Qizong
1726, Number 6, Xiamen	Li Daiheng	
1727, Number 33, Xiamen	Li Daiheng	
1730, Number 12, Xiamen	Li Hongzhong 李弘中	In place of Li Daiheng
1731, Number 27, Nanjing	Li Daiheng Huang Ziyu 黄子欲	In place of Huang Xiangwan 黄享萬
1733, Number 11, Xiamen	Li Hongzhong	In place of Li Daiheng

Based on these records, there is a very high probability that Li Daiheng came from Xiamen.

As a result of Yoshimune's efforts, by the late Edo period the state of domestic sugar supply had changed significantly. Kitagawa Morisada 喜多川守貞 recorded historical rumors and popular knowledge in *Morisada mankō* 守貞漫稿 and *Ruijū kinsei fūzoku shi* 類従近世風俗志, composed between 1837 and 1853. According to volume 28, "Foodstuffs 食類," sugar imported on Dutch vessels was called "Dejima White (*Dejimajiro* 出島白)" after the Dutch trading post at Dejima. Sugar imported from China fell into three categories: highest quality "*sanbon* 三盆", followed by "*jōhaku* 上白," and lowest-quality "*taihaku* 太白".[67] Sugar production in Japan had also spread, with Suruga and Enshū 遠州 growing sugarcane. By the Tenmei 天明 and Kansei 寛政 period (1781–1800), cultivation had spread to Shikoku. By the mid-nineteenth century, Sanshū 讃州 in Shikoku had become the largest producer, followed by Awa 阿波. Suruga, Tōtōmi 遠江, Mikawa 三河 and Senshū 泉州 also produced

65 *Ka-I hentai* ge, 2984.

66 *Tōsen shinkō kaitōroku*, 74, 78, 82, 86, 90, 93, 96.

67 Kitagawa, *Ruijū kinsei fūzoku shi*, 441.

sugarcane. Muscovado was transported from Satsuma, but most was imported from Ryūkyū. Domestically, Kishū 紀州 and Doshū 土州 produced the greatest amount of muscovado, with Senshū, Suruga, Tōtōmi and Mikawa following. By the mid- and late Edo periods, sugar was used not only to make candied sweets, but also in a wide range of food products and cooking, including soba, tempura and fish sausage.

In 1823, Matsura Seizan 松浦静山 (1760–1841) was assigned to Suruga Castle, where he observed the state of sugarcane harvesting and sugar refinement. He recorded how the sugarcane harvest relied on oxen to liquefy the sugarcane, which was then used to make sugar.[68] Matsura also described how Tokugawa Yoshimune had championed the spread of sugar, and claimed that sugar's spread was a major contribution of Yoshimune's reign.[69] Seizan started chronicling events of his time in the *Kasshi yawa* 甲子夜話 in 1821, suggesting that by the middle of the century domestic sugar production had become a widely known fact. In other words, in the nearly one hundred years between the Kyōhō years (1716–1735) under Yoshimune (1684–1751) and the period when Matsura was writing *Kasshi yawa*, sugar production from sugarcane had become common across Japan.

A similar picture of the era emerges from two representative, well-known authors of agricultural texts. Miyazaki Yasusada 宮崎安貞 (1623–1697), an agronomist of the early to mid-Edo period, wrote *Nōgyō zensho* 農業全書 in 1697. In volume 5, "Plants of the countryside: sugarcane (山野菜之類：甘蔗)," Miyazaki wrote that sugarcane would not grow except in hot regions. In the Genroku period (1688–1703) sugarcane seedlings were imported from Ryūkyū to Satsuma, and grown there. However the prohibitive cost required meant that cultivation by individuals was difficult, especially if powerful men such as the local lord did not contribute.[70]

Yet by the time Ōkura Nagatsune 大蔵永常 (1768–1861), an agronomist of the late Edo era, wrote his *Kōeki kokusan kō* 広益国産考 in 1859, the situation had changed drastically. In the second volume, "Sugar 砂糖の事," Ōkura wrote:

> Until just over 200 years ago, only noble people knew about sugar, and those of humble birth never saw it. Yet from the Genroku period through the An'ei (1772–1780) and Kansei periods (1789–1800), a type of muscovado called "Chinese black 唐黒" was imported on Chinese ships. In the Bunka

68 Matsura, *Kasshi yawa* 4, 78.

69 Matsura, *Kasshi yawa* 3, 295.

70 Miyazaki, *Nōgyō zensho*, 391–92.

> period (1804–1817) this ended and no more was imported. Compared to muscovado used today the grains were finer, and of higher quality.[71]

The role of sugar in Japan changed significantly over the two hundred years before Ōkura's 1859 *Kōeki kokusan kō*, dating back roughly to when Miyazaki Yasusada wrote *Nōgyō zensho*. As Ōkura noted, "until just over two hundred years ago, only noble people knew about sugar and those of humble birth never saw it," suggesting that the nobility, and particularly the residents of Edo Castle, were likely to be the greatest consumers of sugar. One study revealed the amount of sugar consumed at Edo Castle to be a surprising one thousand *kin* per day;[72] putting aside the question of whether this number was accurate, it is not difficult to imagine that Edo Castle would have consumed more sugar than anywhere else. Although in the Genroku period only noble people encountered sugar, Ōkura Nagatsune's account shows that over the subsequent two hundred years domestically produced sugar became widespread. In particular, domestic producers began to make sugar equal in quality to that imported on Chinese merchant ships.

4.2 *Expansion of Sugar Consumption in Japan*

By the mid- to late-Edo period, the amount of sugar consumed and its uses had expanded considerably. One example of this spread lies in the use of sugar in iced drinks, sold by "iced water vendors" (*hiyamizuuri* 冷水売) in the Edo era, as described in *Morisada mankō* 守貞漫稿, volume 5, "Occupations 生業." In particular, in Kyoto and Osaka the job was known as "sugar water vendor" (*satōmizuuri* 砂糖水売).[73] Furthermore, candy known as "sugared beans" (*satōiri kintoki* 砂糖入金時)[74] and boiled dumplings made from rice called "rice flour dumplings" (*shiratamauri* 白玉)[75] were both sold with white sugar sprinkled on top. For the annual Boys' Festival, held on the fifth day of the fifth month, bean jam rice cakes (*kashiwamochi* 柏餅) were sold in Edo, Kyoto and Osaka, and contained red bean jam made with sugar. Some cakes in Edo used soybean paste (*miso* 味噌) made with sugar.[76] Children's sweets in the shape of bonito flakes (*katsuobushi* 鰹節) were made by knead-

71 Ōkura, *Kōeki kokusan kō*, 98–101.

72 Kitamura, *Samidare zōshi*, 17; see also Asakura, "Kōki", in vol. 3, 355.

73 Kitagawa, *Ruijū kinsei fūzoku shi*, 174.

74 Ibid., 177.

75 Ibid., 177.

76 Ibid., 270.

ing flour with sugar syrup, and hardening the mixture.[77] The well-known *Kinzanji miso* 金山寺味噌 was made with soybeans and malt barley, to which sugar was added.[78] *Kinton* 金團 was also made with sugar.[79] The bean paste used as a sauce for *konnyaku no dengaku* 菎蒻の田楽 (*konnyaku* on skewers) was sweetened with sugar.[80]

The well-known "*konpeitō*" 金平糖, a small candy made by crystallizing sugar around a poppy seed core and first imported during the Azuchi Momoyama period (c.1573–1600), could not be made without sugar.[81] Sugar also became indispensable to jellied bean paste (*neriyōkan* 練り羊羹).[82] Moreover, steamed buns (*manjū* 饅頭) included sugar.[83] Naturally, expensive *manjū* used expensive sugar. By the Bunka period (1804–1818) *manjū* made with sugar had spread not only to the big cities such as Edo or Osaka, but also to provincial towns.[84] Although the bun of a *soba manjū* was made with buckwheat flour, the sweet jam filling incorporated high-grade imported sugar and *azuki* 小豆 beans.[85] This variety of *manjū* was quite popular in Edo.

In addition to these various sweets and food products, sugar was used in a variety of preserved foods. The Bunka-period *Bōkaen manroku* 卯花園漫録 (volume 2), contains records of plum wine and tangerines (*mikan* 蜜柑) preserved in sugar.[86] Goods such as plum wine and preserved tangerines, commonly used to this day, relatively regularly incorporated sugar. These records also appear in the late Tokugawa nativist scholar Oyamada Tomokiyo's 小山田與清 (1783–1847) chronicle of events from the Bunsei years (1818–1830) onward, the *Matsuya hikki* (松屋筆記, volume 74), where the procedures for "Making plum wine" and "Making candied goods" are described in nearly identical language as the *Bōkaen manroku*.[87]

Takizawa Bakin 滝沢馬琴 (1767–1848) recorded the state of domestic sugar prices in the Tenpō years (1830–1843) in *Ibun zakkō* 異聞雑稿, volume 1.[88] He described a rapid inflation in the spring of 1833, which drove up prices of

77 Ibid., 313.
78 Ibid., 418.
79 Ibid., 433.
80 Ibid., 435.
81 Ibid., 439–40.
82 Ibid., 443.
83 Ibid., 444.
84 Ibid., 445.
85 Ibid., 446.
86 Isonokami, *Bōkaen manroku*, 154–55. See also Asakura, "Afterword" in vol. 5, 409.
87 Oyamada, *Matsuya hikki* 2, 124–25; also Ichijima, "Reigen", 1–5.
88 Takizawa, *Ibun zakkō* jō, 228. See also Asakura, "Afterword," in vol. 2, 434.

white sugar by a factor of 1.9, from 180 *mon* 文 to 350 *mon* per *kin*. Muscovado increased by a factor of 2.4, from 116 *mon* to 280 *mon* per *kin*. This drastic inflation curtailed sugar sales in the countryside, and by the tenth month of that year prices began to fall. Takizawa recorded that white sugar fell to 200 or 180 *mon*, and muscovado to 132 *mon*. In addition to sugar, *mochi* rice also suffered high inflation, causing the size and sweetness of various sweets to diminish. Even after prices fell, sweets remained half the size as before the inflation, though their sweetness did increase somewhat. Similar episodes appear in the records of other periods.

5 Conclusion

From the early Edo period onward, Japan imported foreign-produced sugar on Chinese and Dutch vessels. Chinese coastal merchants transported sugar produced in Xiamen and Quanzhou in Fujian Province and Chaozhou in Guangdong Province to Zhapu in Zhejiang Province. There it was transferred to trading ships, known as *tōsen*, which plied Nagasaki. From Nagasaki it made its way to the merchant houses of Osaka, which distributed Chinese-produced sugar across Japan.

While the import of Chinese sugar continued throughout the Edo period, efforts to increase the volume of domestic sugar production spread across Japan from the Kyōhō years onward. Under the eighth Tokugawa shogun, Yoshimune, Chinese methods of sugarcane cultivation and sugar production were studied through books and also directly from the captains of Chinese ships calling at Nagasaki. Thereafter the *bakufu* urged sugarcane cultivation on shogunal lands and Tokugawa family lands in Suruga, Mikawa, Nagasaki and Kishū, contributing to the full-fledged development of sugarcane production. As a result, by the Bunsei period one hundred years later, Matsura Seizan, head of the Hirado domain, could describe the widespread cultivation of sugarcane and the production of sugar all over Japan in *Kasshi yawa*. Commoners in Edo turned that sugar into a variety of foodstuffs in their diet. Innumerable goods relied upon sugar, including sugar water drinks, rice flour dumplings, bean jam rice cakes, bonito flakes, *Kinzanji miso*, jellied bean paste, *manjū, soba manjū*, and more. By the end of the Edo period, Japanese people used large amounts of sugar on a daily basis, and many goods common then have persisted virtually unchanged to the present day.

CHAPTER 8

Imports and Exports of Books by Chinese Junks in the Edo Period

Matsuura Akira 松浦章

1 Introduction

The Edo Period was the period of Japan's so-called "seclusion policy" (*sakoku* 鎖国), during which foreign trade was limited to Chinese and Dutch merchant ships. Therefore most of the foreign books that arrived from abroad were Western books brought by the Dutch and Chinese books brought by the Chinese. Historian Ōba Osamu 大庭脩 made especially clear in *Edojidai ni okeru Tōsen mochiwatashisho no kenkyū* 江戸時代における唐船持渡書の研究 [Research on the cargo lists of Chinese trading vessels in the Edo period][1] that a large number of Chinese books were imported annually by Chinese junks.

It has long been known that the Japanese enjoyed Chinese books; their significance at the time is simply but vividly recorded in the *Wohao*[2] 倭好 (in Japanese *Wakō*). The *Wohao* reports that, of the Five Classics, *the Classic of History* (*Shujing* 書経 in Japanese *Shokyō*[3]) and *the Classic of Rites* (*Liji* 礼記 in Japanese *Raiki*) were well respected, whereas *the Classic of Changes* (*Yijing* 易経 in Japanese *Ekikyō*), the Classic of Poetry (*Shijing* 詩経 in Japanese *Shikyō*) and *the Spring and Autumn Annals* (*Chunqiu* 春秋 in Japanese *Shunjū*) were considered to be less important. As to the Four Books, we are told that *the Analects* (*Lunyu* 論語 in Japanese *Rongo*), *Great Learning* (*Dazue* 大学 in Japanese *Daigaku*) and *Doctrine of the Mean* (*Zhongyong* 中庸 in Japanese *Chūyō*) were well respected, whereas *the Mencius* (*Mengzi* 孟子 in Japanese *Mōshi*) was less well regarded. It is thought that *the Mencius* was particularly unpopular due to the spread of its reputation as a revolutionary text which

1 Ōba, *Edo jidai ni okeru Tōsen mochiwatashisho no kenkyū*; Ōba, *Edo jidai ni okeru chūgoku bunka juyō no kenkyū*.

2 *Riben-fengtu-ji*, page unknown.

3 Where Chinese book titles have a well-known English translation, the book is referred to using that translation. Otherwise book titles are in the original language. The English translation of these titles is in [ ].

 | DOI 10.1163/9789004304512_009

went against the Shogunate's policy, an idea that is clearly represented in the *Wohao*. Buddhist and Taoist scriptures were also popular, and medical texts, too, were inevitably purchased. In his *Xiuhaibian* 袖海編, Wang Peng 汪鵬, who traded in Nagasaki at the beginning of the Meiwa period 明和年間 between 1772 and 1780, wrote that the *dongren* (東人 Easterners), as he called the Japanese, would purchase any Chinese book imported to Japan however high the price.[4]

Edo-period Sino-Japanese exchange was done by Chinese junks visiting Nagasaki, the only port at which international trade was permitted, from the middle of the 17th century until the 1860s.[5] Chinese goods such as silk, sugar and medicine were brought to Japan by these junks, which returned to China carrying Japanese goods such as copper, sea cucumber, and other dried goods and products of the sea.[6] As already indicated, a great many books were also among the goods brought to Japan from China. Because these were very expensive, they were primarily purchased for the collections of wealthy individuals such as the *shōgun* 将軍 (feudal military administrator) and various *daimyō* 大名 (feudal lords).[7] In return, as it were, older books that had originally been published in China and subsequently reproduced using Japanese wood-block printing were taken back to their original home by Qing Dynasty sailing vessels. This was made possible by a man named Wang Zhuli 汪竹里, the alternate name of the above-mentioned Wang Peng, who visited Nagasaki every year from 1772 to 1780, in the middle of the Qianlong Dynasty. He published *Xiuhaibian* on the occasion of the Chrysanthemum Festival in 1764. It can be called the Edo period's only account of Japan by a Chinese visitor to Nagasaki. Because Wang Peng used the name Wang Zhuli in his business dealings, few realized that they were one and the same person. The Japanese-reproduced books that Wang Peng brought back to China had more than a little influence on the academic community of the Qing Dynasty.[8] Some of these were included in the *Zhibuzuzhai-congshu* 知不足齋叢書 (in Japanese *Chifusokusai-sōsho*), which, together with *Yueyatang-congshu* 粤雅堂叢書 (in Japanese *Etsugadō sōsho*), is considered outstanding for the precision of its editing and the beauty of its printing in the Qing period.[9]

---

4 Matsuura, *Edo jidai Tōsen ni yoru Nicchū bunka kōryū*; *Shōdai sōsho zokuhen*, issue 219, page unknown.

5 Matsuura, *Edo jidai Tōsen ni yoru Nicchū bunka kōryū*, 346–48.

6 Matsuura, *Shindai kaigai bōekishi no kenkyū*, 382–402.

7 Ōba, *Edo jidai ni okeru Chūgoku bunka juyō no kenkyū*, 128–29, 132–33, 212.

8 Matsuura, *Edo jidai Tōsen ni yoru Nicchū bunka kōryū*, 202–16.

9 Momose, "Chifusokusai sōsho", 157.

This paper has thus far described how Chinese books were brought to Nagasaki by Chinese junks during the Edo period, how books taken to China from Nagasaki were included in Bao Tinbo's 鮑廷博 *Zhibuzuzhai-congshu*, and the influence that books printed in Japan had on Qing Dynasty intellectuals. In the following sections, it will explore in more detail the impact that books traded across the sea had on Qing China and Tokugawa Japan.

2 Books Brought by Chinese Junks and Reproductions Printed by Edo Booksellers

Neo-Confucianism (*Zhuzixue* 朱子学, in Japanese *Shushigaku*) was the dominant school of thought among Edo-Period academics, and it can be said that the Confucian classics and Buddhist texts were the dominant printed works.[10] Of all Chinese book imports, the most popular were on scientific theories from the Ming Dynasty through to the beginning of the Qing Dynasty. Many Ming and Qing Dynasty works were imported, ranging from collections of poetry to *kogan* 古玩 (old books that were particularly rare or valuable), *sekifu* 石譜 (genealogies and suchlike, printed using a stone press), music and novels.[11]

Some Chinese junks carried a large quantity of books. For example, in 1753, the wreck of the *inuban gaisen* 戌番外船[12] drifted ashore at Hachijō-jima 八丈島, carrying 441 types of books, 490 copies in 1,476 sets, comprising 12,082 books in all.[13]

As an example of the books printed by booksellers in the Edo period, let us consider a catalogue of Chinese books published in 1798 as *Seireikaku zōhan shomokuroku* 青藜閣蔵板書目録 [Catalog of printing blocks owned by the Seireikaku bookshop]. This bookshop was owned by Suharaya Ihachi 須原屋伊八 and located at Asakusa Kayachō nichōme in Edo.

- *Sishu-jizhu* 四書集註 [a collection of Zhu Xi's commentated versions of *the Analects, the Mencius, Great Learning* and *Doctrine of the Mean*]. Phonetic readings (of the Chinese text) by Hayashi Dōshun 林道春.[14] Published by this bookshop. Ten volumes.

---

10 Yayoshi, *Mikan shiryō ni yoru Nihon shuppan bunka*, 119.

11 Ibid., 120–21.

12 The direct translation is "extra dog ship." The name refers to the fact that the ship arrived in the Year of the Dog, but it was not part of the port's regular traffic.

13 Ōba, *Hōreki sannen hachijōjima hyōchaku Nankinsen shiryō*, 469.

14 Hayashi Dōshun is the Buddhist priest name Hayashi Razan 林羅山.

- *Sishu-jizhu* (a different printing of the same text), proofread by Zhu Xiqi 朱錫旂, additional annotations by Satō Issai 佐藤一齋. Ten volumes.
- *Zhouyiguzhu* 周易古註 [an annotated text on *the Classic of Changes*, one of the Five Classics]. Annotated by Wang Bi 王弼 of Wei 魏 and Han Kangbo 韓康伯 of Jin 晉. Five volumes.
- Original text of the Classic of the Changes. Proofread by Oda Kokuzan 小田穀山. Three volumes.
- *Wulixiaoshi* 物理小識 (in Japanese *Butsuri shōshiki*) [An encyclopedia of the science and art of the early Qing Dynasty compiled by Fang Yizhi 方以智]. By Fang Mizhi 方蜜之[15] of the Ming Dynasty. Twelve volumes.
- *Wenzi-quanshu* 文子全書 (in Japanese *Bunshi zensho*) [Works relating to Taoism]. Proofread by Irie Nanmei 入江南溟. Four volumes.
- *Qijin Mengzi-Kaowen-Buyi* 七經孟子考文補遣. [A text revised by editing the portions of *the Mencius* added to the seven texts that form the cornerstone of Confucianism: *The Classic of Changes, the Classic of History, the Classic of Poetry, the Classic of Rites, the Spring and Autumn Annals, the Analects*, and *the Classic of Filial Piety* (*Kōkyō* 孝經)]. Compiled by Yamai Tei 山井鼎. Supplement by Butsu Shukutatsu 物叔達.[16] 32 volumes.
- *Yantie-lun* 鹽鐵論 (in Japanese *Entetsuron*) [Text of pro and contra arguments regarding the monopolies on salt, iron and alcohol under Emperor Wu in the early Han Dynasty]. Compiled by Huan Kuan 桓寬 of the early Han Dynasty. Proofread by Cheng Rong 程栄 of the Ming Dynasty. Six volumes.
- *Zhuzi Xinxue lu* 朱子心學録 (in Japanese *Shushi shingakuroku*) [A text on the Neo-Confucianism of Zhuzi 朱熹]. Compiled by Jinxi Wangyi 金谿王萓 of the Ming Dynasty. Two volumes.
- *Daxue Zhangju Zuanshi* 大学章句纂釋 (in Japanese *Daigaku shōku senshaku*) [A commentary on *Great Learning*] and *Zhushuo Bianwu* 諸説辨誤 [A pro and contra investigation of various theories]. By Koga Seiri 古賀精里. Two volumes.
- *Sishu Bianjiang* 四書便講 (in Japanese *Shisho benkō*) [A text on *Great Learning, Doctrine of the Mean, the Analects* and *the Mencius*]. By Satō Naokata 佐藤直方. Six volumes.
- *Nüjie* 女誡 (in Japanese *Jokai*) [A book on the behavior of women], by Cao Shishu 曹世叔 of the late Han Dynasty. Newly revised by Chū Chin 忠珍. One volume.

15 Fang Mizhi is the pen name (*azana* 字) of Fang Yizhi.

16 He is also known by the pen name Ogyū Hokkei 荻生北渓. Shukutatsu was a pen name, and Butsu was his Chinese surname. His given name (*na* 名) was Kan 観.

- *Tang-yin-bi-shi* 棠陰比事 (in Japanese *Tōin hiji*) [A text consisting of records of trials (interrogations and the passing of sentences) in ancient China, gathered to be used as a reference in trials]. By Gui Wangrong 桂萬栄[17] of the Ming Dynasty. Three volumes.
- *Xiaojing Huitong* 孝經會通 [An annotated text of the Classic of Filial Piety, considered a work of Confucius, in which he explains the way of filial piety, by which children pay respect to their ancestors and parents]. By Shen Huai 沈淮 of the Ming Dynasty. Proofread by Asakawa Zen'an 朝川善庵. One volume.
- *Xizhao-leshi* 熙朝樂事. By Tian Rucheng 田汝成 of the Ming Dynasty. Translated by Owari Nankō 尾張南康. One volume.
- *Song-shi-chao* 宋詩鈔 (in Japanese *Sōshishō*) [A representative collection of Song Dynasty poetry]. Compiled by Zhang Erming 張二銘 of the Qing Dynasty. Revised by Yamamoto Hokuzan 山本北山. Four volumes.
- *Jin-shi-chao* 金詩鈔 (in Japanese *Kinshishō*) [A representative collection of Jin Dynasty poetry]. Compiled by Gu Kuiguang 顧奎光 of the Qing Dynasty. Revised by Tate Ryūwan 館柳彎. Four volumes.
- *Yuan-shi-chao* 元詩鈔 (in Japanese *Genshishō*) [A representative collection of Yuan Dynasty poetry]. Compiled by Zhang Erming of the Qing Dynasty. Revised by Satō Issai. Four volumes.
- *Fang-qiu-ya-shi-chao* 方秋崖詩鈔 [A representative collection of the poetry of Fang Yue 方岳[18]]. Compiled by Wu Mengju 呉孟擧 of the Qing Dynasty. Proofread by Namikawa Tenmin 並河天民 and Sawa Tansai 佐羽淡齋. Two volumes.
- *Leng-zhai-shi-hua* 冷齋詩話 (in Japanese *Reisai shiwa*) [A collection of poetry]. By Seng Huihong 僧惠洪 (a Song Dynasty Buddhist priest). Two volumes.
- *Sui-yuan-shi-chao* 随園詩鈔 (in Japanese *Zuien shishō*) [A collection of poetry taken from a representative selection of the poetry of Yuan Mei 袁枚]. Compiled by Ichikawa Kansai 市河寛斎. Three volumes.
- *Quan-Tang-shi-yi* 全唐詩逸 (in Japanese *Zentō shiitsu*) (A collection of Tang Dynasty poetry that had made its way to Japan). Compiled and edited by Ichikawa Kansai. One volume.

17 Actually, Gui Wangrong lived during the Song Dynasty, and this book was revised by Wu Ne 呉訥 of the Ming Dynasty.

18 Fang Yue wrote under the pen name Qiuya 秋崖.

- *Xi-yu-wen-jian-lu* 西域聞見録 (in Japanese *Saiiki bunkenroku*) (A Qing Dynasty text on the geography of the northwest). By Qishiyi 七十一[19] of the Qing Dynasty. Phonetic readings by Dōun 道雲. Three volumes.
- *Ji-xiao-xin-shu* 紀効新書 (in Japanese *Kikō shinsho*) [A text on military strategy]. By Qi Nantang 戚南塘 of the Ming Dynasty. Revisions by Hei Shiryū 平子龍.[20] Six volumes.
- *Xi-yang-huo-gong-shen-qi-shuo* 西洋火攻神器説 (in Japanese *Seiyō kakō shinkisetsu*) [A text on Western firearms and their military use]. By He Rubin 何汝賓 of the Ming Dynasty. Revised by Hirayama Heigen.
- *Xi-yang-huo-gong-shen-qi-shuo* (a different printing of the same text), annotated with Japanese pronunciations and simple explanations.

From the above, one can see the sorts of Chinese books that were popular in Japan. Due to the low number of Chinese imports in circulation, many Japanese-made reproductions were published; a case in point is *Sui-yuan-shi-chao* [Selected poems of Sui-yuan], one of the books contained in the above list.

It is well known that Sui-yuan 随園 was the name of a park in an outlying district of Nanjing 南京 in which the famous poet Yuan Mei (1716–1797) resided. But when did his poetry find its way to Japan? Let us consider some of the shipping records included in Ōba Osamu's *Edojidai ni okeru Tōsen mochiwatashisho no kenkyū*.

- 1791: One bound copy of *Sui-yuan-shi-hua* 隨園詩話 (in Japanese *Zuien shiwa*).[21] Listed in the *Shōhaku sairai shomoku* 商舶載來書目, a catalog of books imported by Chinese trading ships, vol. "su".
- 1793: One bound copy of *Xiaocang-shan-fang* 小倉山房 (in Japanese *shōsō sanbō*).[22] Listed in the *Shōhaku sairai shomoku*, vol. "se".[23]
- 1794: 15 bound copies of *Xiaocang-shan-fang*. Listed in *Kansei rokunen tora niban Nankinsen shoseki meimoku* 寛政六年寅貳番南京船書籍名

19 His pen name is Chunyuan 椿園.

20 This is the pen name of Hirayama Heigen 平山兵原, a vassal and military tactician.

21 Ōba, *Edojidai ni okeru Tōsen mochiwatashisho no kenkyū*, 739.

22 Xiaocang-shan is the name of a mountain in an outlying district north of Nanjing, on which Yuan Mei built a garden attached to a residence, which he named *Xiaocang-shan-fang*, literally "the room at Xiaocang-shan," a name which he also gave to many of the works he published.

23 Ōba, *Edojidai ni okeru Tōsen mochiwatashisho no kenkyū*, 733.

[a catalog of books shipped from Nanjing on the second ship that arrived in 1795].[24]

- 1798: One bound copy of *Xiaocang-shan-fang chidu* 小倉山房尺牘 (in Japanese *Shōsō sanbō sekitoku*) [a collection of Yuan Mei's correspondence]. Listed in *Shōhaku sairai shomoku*, vol. "se".[25]
- 1798: One bound copy of *Xiaocang-shan-fang wen-chao* 小倉山房文鈔 (in Japanese *Shōsō sanbō bunshō*) [Excerpts from Yuan Mei's works]. Listed in *Shōhaku sairai shomoku*, vol. "se".[26]

It is thought that the first of Yuan Mei's works to be shipped to Japan was *Sui-yuan-shi-hua*, probably the edition published in 1790. If this is the case, then it was shipped to Nagasaki the year after its publication.

According to *Sui-yuan-shi-chao fanli* 随園詩鈔凡例 (in Japanese *Zuien shishō hanrei*) [Explanatory notes to *Sui-yuan-shi-chao*], written in 1815/5 by Ichikawa Kansai (1749–1820), twenty years had already past since the arrival of *Sui-yuan-shi-hua*, and it had become much esteemed by the literati. However, because few copies had been imported, there were not so many people who recognized the significance of Yuan Mei's works.[27] Twenty years before 1815 is 1795. This is near enough to the year 1791 in which *Sui-yuan-shi-hua* arrived in Nagasaki according to Oba's research. Thus it is conceivable that *Sui-yuan-shi-hua* first arrived in Japan in 1791.

In Nagasaki, Kansai first acquired the 31-volume *Xiaocang-shan-fang shichao* 小倉山房詩鈔 (in Japanese *Shōsō sanbō shishō*) [Poetry taken from *Xiaocang-shan-fang*], which contained over 1,500 poems. He subsequently purchased *Xiaocang-shan-fang quanji* 小倉山房全集 (in Japanese *Shōsō sanbō zenshū*) [The complete *Xiaocang-shan-fang*], and in later years recorded, in his own 1815 publication *Sui-yuan-shi-chao fanli* [Explanatory notes to *Sui-yuan-shi-chao*], his amazement at the number of poems in that 37-volume collection.[28]

In *Zuien shishō jo* 随園詩鈔序 [Introduction to *Sui-yuan-shi-chao*], Ōkubo Kōna 大窪行, who learned Chinese poetry with Kansai and wrote under the pen name Ōkubo Shibutsu 大窪詩仏,[29] wrote the following:

24 Ibid., 251.

25 Ibid., 738.

26 Ibid., 738.

27 Nagasawa, *Wakokubon kanshi shūsei*, 192.

28 Ibid., 193.

29 Rai, "Ōkubo Shibutsu", 543.

> In literature, all kinds of happy coincidences occur, but in one instance these coincidences conspired in a singular manner. Some twenty years ago, Ichikawa Kansai opened the private school *Kōkosha* 江湖社, where he taught the poetry of such authors as Bai Letian 白楽天, Li Bai 李白 and Wang Wei 王維. In addition, he recited the refreshing poetic works of Yuan Mei's *Xiaocang-shan-fang*. My only regret is that I was not able to see Yuan Mei's complete works. However, in 1813, Ichikawa Kansai travelled to Nagasaki, where he was able to acquire a copy of *Cang-shan ji* 倉山集 (in Japanese *Sōsan shū*), from which he shall publish a selection for a number of intellectuals. He has asked me to write the introduction for that publication, which task I have hereby accomplished—Ōkubo Shibutsu.[30]

It thus appears clear, from both this text and Kansai's *Zuien shishō hanrei*, that Kansai travelled to Nagasaki in the course of his duties, and there acquired *Cang-shan ji*, i.e., the complete version of *Xiaocang-shan-fang shiji* 小倉山房詩集.

In *Ichikawa Shisei boketsumei* 市河子静墓碣銘 [epitaph for Ichikawa] in the third volume of *Jijitsu bunpen* 事実文編 [a biographical reference focused on famous intellectuals of the Edo period], Hayashi Taira 林衡 (the posthumous name of Hayashi Jussai 林述齋, 1768–1841) writes that Kansai travelled to Nagasaki.[31] The first volume of Kansai's *Keiho muyoroku* 瓊浦夢餘録 [Impressions of Nagasaki] is a record of one of his trips to Nagasaki. In it, he writes that he left in 1813/7 to accompany Makino Yamatonokami Shigetaka 牧野大和守成傑,[32] who had been appointed magistrate of Nagasaki, and that they arrived at the magistrate's office at Inasayama 稲佐山 in Nagasaki on 1813/9/7.[33]

In the winter section (Kiyū fuyu no jō 癸酉冬の条) of *Keiho Muyoroku*, Kansai wrote a poem in which he mourned the death of Yuan Mei while reading *Xiaocang-shan-fang*.[34] From it, one can surmise that Kansai experienced an extraordinary longing for Yuan Mei while reading the complete version of *Xiaocang-shan-fang* that had come into his possession in Nagasaki.

The poems of Yuan Mei's *Xiaocang-shan-fang* are arranged by time period. Based on that arrangement, let us specifically compare the differences between

30 Nagasawa, *Wakokubon kanshi shūsei*, 191.

31 Gokyū, *Jijitsu bunpen*, 307.

32 Makino Yamatonokami Shigetaka 牧野大和守成傑 held the position of magistrate from 1813 to 1815. See *Zoku Nagasaki jitsuroku taisei*, 3.

33 Ichikawa, *Kansai yokō*, 265.

34 Ibid., 295.

the selected works of *Xiaocang-shan-fang* that Kansai saw with the complete poetry collection. The following table compares the number of poems from each section of the complete collection and establishes how many of those poems are included in the selected works.

TABLE 8.1 *Comparison of the complete* Xiaocang-shan-fang (Xiaocang-shan-fang shiji 小倉山房詩集) *and the collection of selected works* (Xiaocang-shan-fang shichao 小倉山房詩鈔) *bought by Ichikawa Kansai*[35]

Volume[36]	Published	Length[37]	Poems in complete collection	Poems in selected works
1	1736, 1737	14 (f.)	52	17
2	1739–1741	14 (f.3)	54	17
3	1742, 1743	19 (f.7)	98	55
4	1744, 1745	14 (r.8)	84	15
5	1746, 1748	25 (f.)	102	32
6	1749	17 (f.1)	72	20
7	1750, 1751	21 (f.1)	90	47
8	1752	28 (7)	149	95
9	1753		72	60
10	1754	26 (f.8)	125	78
11	1755	11 (r.8)	94	33
12	1756	11 (r.10)	31	21
13	1757	22 (r.)	102	75

35 The comparisons were made using books from the Shanghai Library 上海図書館. The complete collection values come from *Xiaocang-shan-fang shiji* 小倉山房詩集 [Collected *Xiaocang-shan-fang* poems], library identification number *zhang* 長 37124–33, dimensions 16.2 cm wide × 24.9 cm high, ten volumes. The selected works values come from *Xiaocang-shan-fang shiji buyi-* 小倉山房詩集補遺 [Collected *Xiaocang-shan-fang* poems, supplement], library identification number *zhang* 長 334533–36, dimensions 11.8 cm wide × 17.5 cm high, four volumes, small-sized version. The latter is clearly entitled *Xiaocang-shan-fang shiji* on its cover, but its title page reads *Xiaocang-shan-fang shichao* [Selected *Xiaocang-shan-fang* poems].

36 Volume (*maki* 巻) here refers to the number of the volume in the complete collection in ten books.

37 The number listed here is the number of leaves the volume takes up (*chōsū* 丁数). An "f" in parentheses indicates the front side of the leaf, and an "r" indicates the reverse. The number following "f" or "r" refers to the number of lines.

TABLE 8.1 *Comparison of the complete* (cont.)

Volume	Published	Length	Poems in complete collection	Poems in selected works
14	1758	14 (r.8)	72	27
15	1759	30 (f.3)	152	98
16	1740,1761	23 (r.1)	102	21
17	1762, 1763	23 (r.10)	133	30
18	1764	16 (r.4)	56	28
19	1765	19 (r,6)	92	28
20	1766, 1767	29 (r.10)	135	80
21	1768, 1769	22 (r.6)	70	70
22	1770, 1771	18 (r.9)	61	24
23	1772, 1773	16 (r.10)	79	29
24	1774, 1775	21 (f.6)	120	26
25	1776, 1777, 1778	37 (f.9)	190	103
26	1779, 1780	35 (r.3)	207	81
27	1781	17 (f.)	118	59
28	1782	41 (r.)	219	70
29	1783	20 (f.8)	101	10
30	1784	56 (r.7)	234	158
31	1785, 1786	11 (f.4)	137	55
32	1787–1789	38 (r.7)	32	
33	1791	16 (f.1)		
34	1793	12 (f.7)		
Total			3,435	1,562

From this it is clear that what Ichikawa Kansai found was a copy of the less comprehensive selected works, which explains the astonishment recorded in *Zuien shishō hanrei* that he could not help but experience when he later discovered an edition of the complete version.

It seems that Kansai, who was enthusiastic about Yuan Mei's *Sui-yuan-shi-hua*, had a strong wish to popularize Yuan Mei's work in Japan. Thus he selected some poems from the selected work *Xiaocang-shan-fang shichao* and published the excerpt as *Zuien shishō* 隨園詩鈔. However, despite the similarity of this title to *Sui-yuan-shi-hua* 隨園詩話, it has nothing to do with *Sui-*

yuan-shi-hua as regards contents. After this publication, Kansai learned of the existence of the complete version of *Xiaocang-shan-fang* and was amazed that many more works were contained in the complete version.

Seireikaku zōsho hansho mokuroku 青藜閣蔵板書目録 [Catalog of printing blocks owned by the Seireikaku bookshop], owned by Suharaya Isaburō 須原屋伊三郎 of Asakusa Sakayachō sanchōme in Edo and inserted in the back of volume three of Ichikawa Beian's 市河米菴 1812/10 publication *Beian bokudan* 米菴墨談, contains the following entry describing Ichikawa Kansai's *Zuien shishō*:

> ***Zuien shishō***—Edited by Ichikawa Kansai, three volumes
> 400 of the 1,500 poems published in *Xiaocang-shan-fang*, selected for easy comprehension for the newcomer to poetry. The author, Zuien (Yuan Mei) is a master poet of the Qianlong period who revolutionized poetry in the early Qing Dynasty. Any student of modern poetry must read this book.

In short, it was published as an essential reader for Edo-period students of Chinese poetry. Its table of contents indicates that it contained 441 poems in total:

- Volume 1: 58 five-line *koshi* 古詩 (ancient poems)
- Volume 2: 19 seven-line *koshi*
- Volume 3: 45 five-line *risshi* 律詩 (ancient Chinese verse form), 2 five-line *hairitsu* 排律 (long version of *risshi*)
- Volume 4: 96 seven-line *risshi*
- Volume 5: 9 five-line zekku 絶句 (a Chinese quatrain), 9 six-line *zekku*
- Volume 6: 203 seven-line *zekku*

As discussed previously, it is certain that Yuan Mei's *Sui-yuan-shi-hua* found its way to Japan some time immediately following publication. Its influence was huge and it was widely known by the Japanese literati. However one may surmise that only few copies of the text were available in Japan. It is reasonable to think that, because of this, these few available copies were lent out to be copied by hand and studied. It is also certain that Ichikawa Kansai, who regretted the rarity of printed books, visited Nagasaki on official business, where he acquired a collection of selected *Xiaocang-shan-fang* poems. He then arranged selected poems from that collection by form, and republished them as *Zuien shishō* 随園詩鈔. Despite their very similar names when written in Chinese characters (only the final character differs), this text was, as previously indicated, different from Yuan Mei's *Sui-yuan-shi-hua* 随園詩話. The complete *Xiaocang-shan-fang* fills 37 volumes, with an additional two supplementary volumes. The collection which Kansai initially acquired, and upon which *Zuien shishō* was based, was a simplified version and contained only 31 volumes. If one considers the state of Sino-Japanese relations at the time, this is an extraordinarily rapid reception of a written work.

3 Export of Chinese Books Published in Japan to China

To ascertain the kind of books published in Japan that were brought from Nagasaki to China on Chinese junks during the Edo period, let us begin with the records contained in *Nagasakishi zokuhen* 長崎志續編 [Sequel on the history of Nagasaki]. This was an Edo-period publication containing a chronological listing of events relating to Nagasaki, the main volume of which was entitled *Nagasakishi* 長崎志 [History of Nagasaki] or *Nagasaki jitsuroku taisei* 長崎實録大成 [Nagasaki complete record of events]. The supplementary volume was called *Zoku Nagasaki jitsuroku taisei* 續長崎實録大成 [Sequel on the Annals of Nagasaki].

The following descriptions are contained in *Tōsen shinkō narabi ni zatsuji no bu* 唐船進港并ニ雜事之部 [Various accounts of events related to the arrival of Chinese junks at port], found in volume eight of *Nagasakishi zokuhen*:

- 1794: The sixth ship of the year 1793 purchased one copy of *Shichikei Mōshi kōbun hoi* 七經孟子考文補遺 [Supplement to the seven Confucian classics and *the Mencius*] published in Japan, before setting sail.[38]
- 1801: Before leaving for China, the merchants of the first and second ships of the year 1801 took on board a total of eight volumes as samples, comprising the first and second volumes of *Yi-cun-cong-shu* 佚存叢書.[39]
- 1809: One each of a number of ancient books published in Japan was included among the volumes loaded on ships returning to China this autumn. These included: *Erya-zhushu* 爾雅註疏, *Yijing-benyi* 易經本義, *Nippon shoki* 日本書記, *Daxue-jie* 大学解, *Zhongyong-jie* 中庸解, *Shujing-guzhu-yinyi* 書經古註音義, *Liji-guzhu* 禮記古註, *Chunqiu-jizhu* 春秋集註, *Gongyang-zhuan* 公羊傳, *Guliang-zhuan* 穀梁傳, *Zuo-zhuan-xida-yilu* 左傳觽大疑録, *Hi-Sorai-gaku* 非徂徠學, *Lunyu-zheng-zhengwen* 論語徵正文, *Daxue-*

38 *Zoku Nagasaki jitsuroku taisei*, 196.

39 Ibid., 200.

zhangju-xinshu 大学章句新疏, *Guwen-Xiaojing-zhengwen* 古文孝經正文, *Xiaojing-dayi* 孝經大義, *Nippon shisen* 日本詩選, *Shishu-guzhuan* 詩書古傳, *Shijing-jizhu* 詩經集註, *Shujing-jizhu* 書經集註, *Yijing-jizhu* 易經集註, *Liji-jizhu* 禮記集註, *Zhao-zhu mengzi* 趙註孟子, *Lunyu* 論語, *Mengzi* 孟子, *Lunyu-zhushu* 論語註疏.[40]

- 1810: One copy of *Kobai enbokufu* 古梅園墨譜[41] exported on ships returning from Nagasaki to China this spring.[42]
- 1811: One copy of *Dongyi-baojian* 東醫寶鑑 published in Japan exported in autumn from Nagasaki on ships returning to China.[43]
- 1817: Thirteen copies of *Qunshu-zhiyao* 群書治要 published in Japan, exported from Nagasaki in spring and autumn on ships returning to China.[44]
- 1818: Two copies of *Seizai sōroku* 聖済総録, published last year by the *Igakukan* 醫學館 (a medical school founded by the Tokugawa Shogunate). The *Igakukan* had them brought to the port by the magistrate of Nagasaki Tsutsui Izuminokami 筒井和泉守, whom they had asked to have the books given to Chinese merchants and shipped to China. Chinese merchants arriving in port this year brought ten bamboo artworks and suchlike in exchange for these books. Izumi no kami then returned to Edo and delivered the books to the Shogunate; the *Igakukan* was also informed.[45]
- 1823: The following books, published in Japan, were exported to China this spring on the fourth ship of the year 1822 by merchant Liu Jingyun 劉景筠: *Yi-cun-cong-shu* 佚存叢書 vol. 1 (two copies), and vol. 2 and 3 (one copy each); two copies of *Qunshu-zhiyao* 群書治要; three copies of *Lunyu-jijie* 論語集解; one copy of *Shiji-pinglin* 史記評林; and one copy of *Kobaien bokufu kōhen* 古梅園墨譜後編.[46]
- 1824: The following books, published in Japan, were exported to China this spring on the fourth ship of the year 1822 by merchants Liu Jingyun and Zhu Kaizhe 朱開折: One copy each of vols. 1, 2, 3 of *Yi-cun-cong-shu*; one copy each of *Qunshu-zhiyao* and *Lushi-chunqiu* 呂氏春秋. In addition, merchants Shen Qiquan 沈綺泉 and Jiang Yunge 江芸閣 took on the seventh

40 Ibid., 211.

41 *Kobai enbokufu* is the name of an ink shop in Nara. This is a text on calligraphy published by Kobaien 古梅園, whose inks were also exported to China.

42 *Zoku Nagasaki jitsuroku taisei*, 213.

43 Ibid., 214.

44 Ibid., 218.

45 Ibid., 219.

46 Ibid., 225.

ship of the year 1822: One copy each of *Sishu-jizhu* and *Shiji-pinglin*, as well as two copies of *Shanhanlun-jiyi* 傷寒論輯義.[47]

- 1825: The following books, published in Japan, were exported to China by merchants Liu Jingyun and Yan Xuefan 顔雪帆 on the fourth ship of the year 1822: One copy each of *Qijin Mengzi-Kaowen-Buyi, Lunyu-zheng* 論語徵, and *Qunshu-zhiyao*, as well as ten copies each of vol. 1, 2, 3 of *Yi-cun-cong-shu*.[48]
- 1837: This April, one copy of the 25-volume revised *Dongyi-baojian* 訂正東醫寶鑑, published in Japan, was exported from Nagasaki to China by merchant Shen Yungu 沈耘穀 on the third ship of the year 1836.[49]

As recorded in the above entries, during the more than forty-year span from 1794 to 1837, Chinese merchants took books printed in Japan back to China on their own initiative, over and above the publications that the *Igakukan* wished to have brought to China. Table 8.2 contains the books of which more than one copy was shipped from Nagasaki to China by Chinese junks during this time.[50]

TABLE 8.2 *Multi-volume texts shipped to China by Chinese junks from 1794 to 1837, grouped by similar or identical books*

Year of departure	Ship[51]	Text
1801	Numbers 1 and 2 of the year 1801	*Yi-cun-cong-shu-qian-bian hou-bian*, eight copies, hand-copied
1823	Number 4 of the year 1822	*Yi-cun-cong-shu*, two copies of vol. 1, one copy each of vols. 2, 3
1824	Number 3 of the year 1823	*Yi-cun-cong-shu*, one copy each of vols. 1, 2, 3
1825	Number 4 of the year 1825	*Yi-cun-cong-shu*, ten copies each of vols. 1, 2, 3

47 Ibid., 225.

48 Ibid., 225.

49 Ibid., 249.

50 Similar books are counted as the same book, for example *Yijing-jzhu* and *Yijing-benyi*.

51 The year by which the ships are identified is the year of their arrival in Nagasaki and can thus differ from the year of their departure.

Year of departure	Ship	Text
1809		*Yijing-jizhu*,[52] one copy
1809		*Yijing-benyi*, one copy
1817		*Qunshu-zhiyao*, 13 copies
1823	Number 4 of the year 1822	*Qunshu-zhiyao*, two copies
1824	Number 3 of the year 1823	*Qunshu-zhiyao*, one copy
1825	Number 4 of the year 1825	*Qunshu-zhiyao*, one copy
1810		*Kobaien bokufu*, one copy
1823	Number 4 of the year 1822	*Kobaien bokufu kōhen*, one copy
1811		*Dongyi-baojian*, one copy
1837	Number 3 of the year 1836	Revised *Dongyi-baojian*,[53] one copy consisting of 25 volumes
1823	Number 4 of the year 1822	*Shiji-pinglin*,[54] one copy
1824	Number 7 of the year 1823	*Shiji-pinglin*, one copy
1794	Number 6 of the year 1793	*Shichikei Mōshi kōbun hoi*,[55] one copy
1825	Number 4 of the year 1825	*Shichikei Mōshi kōbun hoi*, one copy
1809		*Chunqiu-jizhu*,[56] one copy
1809		*Chunqiu-jizhu*, one copy
1809		*Shujing-guzhu-yinyi*, one copy

52 Possibly the *Yijing-jizhu* published in 1663. See Naikaku bunko, *Kaitei naikaku bunko kanseki bunrui mokuroku*, 2.

53 This is possibly the 1724 edition compiled by Heo Jun 許浚 of Korea. See Naikaku bunko, *Kaitei naikaku bunko kanseki bunrui mokuroku*, 232.

54 A Ming Dynasty text edited by Ling Zhilong 凌稚隆 with supplement by Li Guangjin 李光縉. The 130-volume Japanese edition with a two-volume main text was published in 1636, 1672, 1674 and 1786. See Naikaku bunko, *Kaitei naikaku bunko kanseki bunrui mokuroku*, 58.

55 This is known to be a 1731 publication of *Qijin Mengzi-Kaowen-Buyi* in 32 books with supplement by *Yamanoi Teisen* 山井鼎撰 and *Ogyū Kan* 荻生観 (also known as *Ogyū Hokkei* 荻生北渓, the younger brother of *Ogyū Sorai* 荻生徂徠). See Naikaku bunko, *Kaitei naikaku bunko kanseki bunrui mokuroku*, 604.

56 A 37-volume publication of *Chunqiu-jizhu*, by Hu Anguo 胡安國 of the Song Dynasty, published in 1663 in Japan. See Naikaku bunko, *Kaitei naikaku bunko kanseki bunrui mokuroku*, 21.

TABLE 8.2 *Multi-volume texts shipped* (cont.)

Year of departure	Ship	Text
1809		*Shujing-jizhu*,[57] one copy
1809		*Daxue-jie*, one copy
1809		*Daxue-zhangju-xinshu*, one copy
1809		*Mengzi*, one copy
1809		*Zhaozhu Mengzi*, one copy
1809		*Liji-guzhu*, one copy
1809		*Liji-guzhu*, one copy
1809		*Lunyu*, one copy
1823	Number 4 of the year 1822	*Lunyu-jijie*, three copies
1809		*Lunyu-zhushu*, one copy
1825	Number 4 of the year 1825	*Rongochō* 論語徵, one copy (a commentary on *the Analects* written by Ogyu Sorai)
1809		*Rongochō seibun* 論語徵正文, one copy (a commentary on *the Analects* written by Ogyu Sorai)

As one can clearly see from this list, the great majority were woodcut reproductions of Chinese books, and Japanese books accounted for only a small share.

Between 1764 and 1837, the book most commonly exported from Nagasaki to China on Chinese trading junks was without question the Japanese printing of *Yi-cun-cong-shu*, a wood-block print edited by Hayashi Taira. 60 books were printed between 1799 and 1810, including the following works:

57 A set of 58 books called *Gokyō jūchō* 五經集註 [Annotated collection of the Five Classics], a compilation of the 24-volume *Zhouyi-zhuanyi* 周易傳義, the 10-volume *Shujing-jizhu*, the 15-volume *Shijing-jizhu*, the 30-volume *Liji-jishou* 禮記集說 and the 37-volume *Chunqiu-jizhuan* 春秋集傳, was published in Japan in 1726. See Naikaku bunko, *Kaitei naikaku bunko kanseki bunrui mokuroku*, 26.

- Collection 1 (*daiicchitsu* 第一帙), Part 1 (*daiissaku* 第一册): *Guwen Xiaojing* 古文孝經 [The text of the Classic of Filial Piety, said to have been discovered in a former residence of Confucius during the rule of the Han Dynasty emperor Wu-di 武帝], with commentary by Kong Anguo 孔安國 of the Han Dynasty. One volume.
- Collection 1, Part 2–6: *Wuxing-dayi* 五行大義 [Texts on the five elements: water, fire, wood, metal and earth]. By Xiao Ji 蕭吉 of the Sui Dynasty. Five volumes.
- Collection 1, Part 7–8: *Chengui* 臣軌 [A text on the correct behavior of the ruled]. By Wuhou 武后 of the Tang Dynasty. Two volumes.
- Collection 1, Part 9: *Leshu-yaolu* 樂書要錄 [A book about the music of Confucian texts], three hitherto existing volumes (Volume 5–7).
- Collection 1, Part 10: *Liangjing-xinji* 兩京新記 [A text about Changan 長安 and Luoyang 洛陽], one hitherto existing volume (Volume 3), by Weishu 韋述 of the Tang Dynasty. *Liqiao-zayong* 李嶠雜詠 [A poetry collection] by Li Qiao 李嶠 of the Tang Dynasty. Two volumes.
- Collection 2, Part 11–12: *Wenguan Cilin* 文館詞林 [A collection of Tang-Dynasty poetry], four hitherto existing volumes, by Xu Jingzong 許敬宗 of the Tang Dynasty.
- Collection 2, Part 13: *Wen-gong-zhu-xiansheng-ganxing-shi* 文公朱先生感興詩 [A collection of poetry by Zhu-xi] one volume, by Cai Mo 蔡模 of the Song Dynasty.
- Collection 2, Part 14–19: *Zhouyi-zhuan* (*Tai-xuan-yizhuan*) 周易傳 (泰軒易傳) [A commentated edition of *the Classic of Changes*], six volumes, by Li Zongzheng 李中正 of the Song Dynasty.
- Collection 2, Part 20: *Zuoshi Mengqiu* 左氏蒙求 [A collection of anecdotes of the ancients] taken from *Chunqiu-zuoshi-zhuan* 春秋左氏傳, with commentary by Zuo Qiuming 左丘明, one volume. By Wu Hualong 吳化龍 of the Yuan Dynasty.
- Collection 3, Part 21–25: *Tang-cai-zi-zhuan* 唐才子傳 [A collection of biographies of Tang Dynasty poets], ten volumes. By Xin Wenfang 辛文房 of the Yuan Dynasty.
- Collection 3, Part 26–30: *Wang-han-lin-ji-zhu Huangdi-bashiyi-nanjing* 王翰林集註黃帝八十一難經 [A medical text devoted to *kanpō* 漢方, a form of traditional Chinese medicine], five volumes, by Qin Yueren 秦越人 from Lu Guo (盧國 the name of a country in ancient times which was located in present-day Hubei province).
- Collection 4, Part 31–33: *Mengqiu* 蒙求 [A collection of anecdotes of the ancients], three volumes, commentated by Li Han 李瀚 of the Tang Dynasty.

- Collection 4, Part 34–40: *Cui-she-ren Yu-tang-leigao* 崔舍人玉堂類藁, 20 volumes, and *Cui-she-ren Xi-yuan-leigao* 崔舍人西垣類藁 [Poetry collection of Cui Dun 崔敦 of the Song Dynasty], 2 volumes and supplement 1.
- Collection 5, Part 41–50: *Zhouyi-xin-jiangyi* 周易新講義 [A commentated edition of *the Classic of Changes*], ten volumes, by Gong Yuan 龔原 of the Song Dynasty.
- Collection 6, Part 51–60: *Jingwensong-gong-ji* 景文宋公集 [A collection of literature by Song Qi 宋祁 of the Song Dynasty].[58]

The entirety of *Yi-cun-cong-shu* had not yet been published in 1801, but one can assume that exports after 1825 included the entire collection.

The 50-volume *Qunshu-zhiyao*, compiled by Wei Zheng 魏徵 by order of the Tang emperor, was reprinted using copper type in 47 books. This edition, known as the Suruga edition (Surugaban 駿河版), was missing volumes 4, 13 and 20 of the original. In 1787, it was printed again in a 47-book edition under the title *Genna chūko katsuji insatu jūkan* 元和中古活字印本重刊 [Genna-period old-style movable type edition] by the *Meirindō* 明倫堂 (Meirin Temple) in the domain of Owari 尾張.[59] It is this edition that is thought to be the one exported from Nagasaki to China.

Zhibuzuzhai-congshu, an anthology of many works published by Bao Tinbo during the Qing Dynasty, includes the books that were reprinted in Japan. The contents shows that the anthology includes the following:

- Collection 1: *Guwen Xiaojing Kong-zhuan* 古文孝經孔傳, one volume. Japanese edition of *Guwen Xiaojing*, published in 1732, the original text of which was lost in China.
- Collection 7: *Lunyu-yishu* 論語義疏, ten volumes. A commentated edition of *the Analects* by Huang Kan 皇侃 of the Southern and Northern Dynasties, which had been lost in China and was published using texts that had made their way to Japan.
- Collection 21: *Xiaojing zhengzhu* 孝經鄭註, one volume. A commentated text on the Classic of Filial Piety, by Zheng Xuan 鄭玄, which had been lost in China and was published using texts that had made their way to Japan.
- Collection 26: *Wuxing-dayi* 五行大義, five volumes, Lost in China and published using texts that had made their way to Japan in ancient times.
- Collection 30: *Quan Tang-shiyi* 全唐詩逸 (in Japanese *Zentō shiitsu*), three volumes. A collection of Tang-Dynasty poetry that was lost in China and

58 Naikaku bunko, *Kaitei naikaku bunko kanseki bunrui mokuroku*, 583.

59 Ibid., 277.

republished using texts that had made their way to Japan and collected by Ichikawa Kansai. Compiled by Nipponka Seinei 日本河世寧 (a penname of Ichikawa Kansai).

The inclusion of books reprinted in Japan in *Zhibuzuzhai-congshu* has to be credited to the trader Wang Zhuli, whose alternate name, as indicated earlier, was Wang Peng.[60] Wang Peng had an extremely close relationship with Wang Danwang 王亶望, who had at one time served as governor of Zhejiang province 浙江巡撫.[61] According to the biography of Wang Danwang (*Wang Danwang zhuan* 王亶望傳) in *Qing-shi-gao* 清史稿 [a biographical history of the Qing Dynasty], vol. 339 *Liezhuan* 列傳 126, his reporting to the Qianlong emperor came under suspicion during a rebellion in Gansu province 甘粛, of which the former was then governor. It came to light that he had accumulated a fortune during his time as governor of Zhejiang province, which was found to amount to more than a million *liang* (両 silver currency) when his family properties were confiscated. Among the confiscated goods were a number of books.[62] The following are listed in *Chao yuanren Zhejiang Xungfu Wang Danwang liuzhe-shiwu-gubian-qingce* 抄原任浙江巡撫王亶望留浙什物估邊清冊 [Book on the market value of written materials in the possession of Wang Danwang in Zhejiang during his tenure as the governor of Zhejiang]:[63]

- 20 sets of *Ershiyi-shi* 二十一史 [History of the Dynasty], 164 books in total, valued at 16 *liang* and 4 *qian* (16 両 4 銭).
- 19 sets of *Ershiyi-shi*. 159 books in total, valued at 15 *liang* and 9 *qian*.
- 21 sets of *Ershiyi-shi*. 177 books in total, valued at 17 *liang* and 7 *qian*.
- 4 sets of *Zhibuzuzhai congshu* 知不足齋叢書.[64] 32 books in total, valued at 2 silver *liang*.
- 4 sets of *Lunyu-jijie-yishu* 論語集解義疏 (in Japanese *Rongo shikkai giso*).[65] Valued at 2 *liang* of silver and 2 *qian*.
- Two sets of *Zhibuzuzhai congshu* 知不足齊叢書.[66] 24 books in total, valued at 3 *liang* of silver.

60 Matsuura, *Edo jidai Tōsen ni yoru nicchū bunka kōryū*, 202–6.
61 Ibid., 206–9.
62 *Qing-shi-gao*, 11074–75.
63 *Qianlong chaocheng bantan wu dangan zhuan*, 1924–35.
64 *Qianlong chaocheng bantan wu dangan zhuanbian*, 1932–33.
65 Ibid., 1933.
66 Ibid., 1934.

This list demonstrates that Wang Danwang owned *Lunyu-jijie-yishu* and Bao Tinbo's *Zhibuzuzhai-congshu*, which Wang Peng had procured in Japan.

The *Zhibuzuzhai-congshu* included 765 volumes comprising 30 series and 198 different kinds of texts. This is borne out by the following quotation, which is taken from *Zi-Zajia* 子雜家, in volume 270 *Jingji-kao* 經籍考 14 of *Xu wenxian tongkao* 續文獻通考, compiled by Liu Jinzao 劉錦藻.

> Bao Tinbo's father moved from Huizhou 徽州, Anhui province 安徽 to a residence named *Zhibuzuzhai* 知不足齋 in Hangzhou 杭州, Zhejiang province. Bao Tinbo was fond of reading, and he would read his father's books and collect old texts. He borrowed books from famous Zhejiang libraries such as *Zhao-shi-xiaoshan-tang* 趙氏小山堂, *Lu-shi-baojing-tang* 盧氏抱經堂, *Wang-shi-zhenqi-tang* 汪氏振綺堂, *Wu-shi-baijing-lou* 吳氏拜經樓 and made revised versions of them, and collected various books.[67]

As described above, many of the books brought to China on Qing-Dynasty junks in the Edo period were Chinese books that had been reproduced in Japan. Because many of these books had been lost in China, the deep interest shown by Qing Dynasty intellectuals is easy to understand. Let us consider this phenomenon taking the example of one of these intellectuals, named Xu Zongyan 許宗彥. Xu Zongyan had already mastered the Classics at the age of nine, and he lived in the Deqing district of Huzhou prefecture in Zhejiang province 浙江省湖州府德清縣, quite close to Zhapu in the Pinghu district of Jiaxing prefecture, also in Zhejiang province 浙江省嘉興府平湖縣乍浦, which was the center of Chinese trade with Japan. Because price was no object for him when acquiring rare books, one may suppose that he was most keen to purchase *Yi-cun-cong-shu* when that book was brought from Nagasaki. Then, he learned of the existence of *Wuxing-dayi*, which was included in *Yi-cun-cong-shu*, and of which he made a revised version in 1804/03. Thus, it is reasonable to assume that Xu Zongyan had a strong interest in a copy of *Yi-cun-cong-shu* which was brought by the first and second ships of the year 1801 from Nagasaki in 1801. Considering the time required for transportation and the revision, this book made its mark in China surprisingly quickly.

67 *Qingchao xu wenxian tongkao*, Kao 考, 10148.

Conclusion

As a case study of cultural exchange across national borders and across the sea, this essay has shed light on the flow of books carried by Chinese junks between Qing Dynasty China and Tokugawa Japan. Although the people of Japan had a great interest in many of the academic texts brought to Japan by Qing Dynasty trading vessels and appreciated their value, their high prices and low supply meant that the circulation of such books was not very extensive. However, Edo period book sellers were able to overcome this problem by printing reproductions. This also had a negative side effect, as seen in the example of Ichikawa Kansai's partial publication of *Sui-yuan-shi-chao* rather than the complete works of Yuan Mei. This was unavoidable given the lateness of the import of Yuan Mei's complete works.

In the other direction, the *Zhibuzuzhai-congshu*, which contained Japanese publications that had been brought to China by Chinese junks, had a huge influence on Qing Dynasty academics.[68] It is certain that this kind of publication, known as *sōsho* 叢書 (collection of written works), which afforded many more people the opportunity to read a work, resulted in the collection being well known among intellectuals. The sudden and unexpected arrival of a Chinese work from Japan, which had long been lost in its homeland, gave the Chinese academic community an opportunity to re-recognize the erudition of their predecessors. One could even speak of an academic Renaissance brought about by the return of this book from Japan to its native home.

As this paper illustrates, trade between Japan and China in the Edo period was by no means limited to economic exchange and material goods. It also included the exchange of cultural and scientific knowledge that cannot be overlooked.

---

68 Matsuura, *Kinsei Higashi Ajia kaiiki no bunka kōshō*, 334–47.

References

1 Primary Sources

1.1 *Japanese Primary Sources*

"Akita-gun Ani-dōzan shidai kikigaki" 秋田郡阿仁銅山次第聞書 [Akita-gun Accounts of the Ani Copper Mine]. In: Akiyama Mitsuharu 秋山光春, *Akiyama Mitsuharu nikki ketsuyō* 秋山光春日記結要 [Summary of Akiyama Mitsuharu Diary], held by Akita-ken kōbunshokan 秋田県公文書館 [Akita Prefecture Archives].

Bunka gannen nedoshichū shuppan hikiawasechō 文化元年子年中　出帆引合帳 [*Transaction Records of Ship Departures in the Fiscal Year of 1804*], zen 135 chō 全135丁 [135 leaves]. Held by Nagasaki rekishi bunka hakubutsukan 長崎歴史文化博物館 [Nagasaki Museum of History and Culture], ref. Watanabe 17–113 図書番号：渡辺 17–113.

"Bunka teibō Tōsen hyōchakuki" 文化丁卯唐船漂着記 [Accounts of Chinese Shipwrecks, 1807]. In *Tsūkō ichiran* 6, 6.

Daiisho 大意書 [Mid-18th century Records of Trading through Nagasaki-kaisho]. *Kinsei shakai keizai sōsho* 近世社会経済叢書 [*Premodern Societal Economy Series*]. vol. 4, edited by Honjō Eijirō 本庄栄治郎 et al. Tokyo: Kuresu shuppan クレス出版, 1989.

Dōdaigin mukimuki uketorishōmon hikae 銅代銀向々請取証文控 [Receipt of Income from Copper Sales Revenues]. Mine-ke shiryō 峰家史料 [Mine Family Document] 110–1, held by Ōmura-City Archives 大村市立史料館.

Dōkata kōyōchō 銅方公用帳 [Public Records on Copper]. Sumitomoke bunsho 住友家文書 [Sumitomo family documents], 14–6-2.

Dōkata kōyōchō niban 銅方公用帳二番 [Public Records on Copper, vol. 2]. Sumitomoke bunsho 住友家文書 [Sumitomo Family Documents], 14-6-1.

[Goyōdō no oboe] "Kyōhō gansaru yori ushidoshi, shokoku dōzan e owariai goyōdō no oboe 享保元申より丑年、諸国銅山江御割合御用銅之覚 [Record of Official Copper Quotas from the Mines from 1716 to 1721]". In *"Dōkaisho gokōyōchō (kyōhō yonen)" hoka dō-bōeki kankei shiryō* 「銅会所御公用帳(享保四年)」ほか銅貿易関係史料 [Public records from the copper office (1719) and other related material about the copper trade] Sumitomo shiryō sōsho 住友史料叢書 vol. 18, edited by Sumitomo shiryōkan 住友史料館. Kyoto: Shibunkan shuppan 思文閣出版, 2003.

"Hakuseki kengi roku" 白石建議六 [Hakuseki's Proposal 6]. In *Arai Hakuseki zenshū* 新井白石全集 [Collection of works by Arai Hakuseki] vol. 6, edited by Ichijima Kenkichi 市島謙吉, 240–243. Tokyo: Yoshikawa Hanshichi 吉川半七, 1907.

Hanai Ikkō 花井一好. "Gendōroku" 減銅録 [An account of Copper Production Decline]. In *Nihon keizai sōsho* 日本経済叢書 [Series on the Economy in Japan] vol. 32, edited by Takimoto Seiichi 瀧本誠一, 373–384. Tokyo: Nihon keizai sōsho kankōkai 日本経済叢書刊行会, 1917.

[Hikaechō] "Shōtoku rokunen saru yongatsu nijūkyūnichi ni okeru Hidanokami-sama ōsetsukerare sōrō tatsudoshi yori saru hitsujidoshi made yonkanen-bun kaichaku-dō no insū narabini nedan ni tsuki dō daigindaka no hikaechō 正徳六年申四月廿九日於飛騨守様被仰付候辰年より去未年迄四ケ年分廻着銅之員数并直段付同代銀高之扣帳 [Public record of the amount and price of copper as well as on copper revenue in the four-year period between 1712 and 1715. The record was requested by Hidanokami on 1716/4/29]". In *"Dōkaisho kōyōchō (kyōhō ni-nen)" hoka dō bōeki kankei shiryō* 「銅会所公用帳(享保二年)」ほか銅貿易関係史料 [Public records from the copper office (1717) and other related material about the copper trade) Sumitomo shiryō sōsho 住友史料叢書 vol. 15, edited by Sumitomo shiryōkan 住友史料館. Kyoto: Shibunkan shuppan 思文閣出版, 2000.

Honda Toshiaki 本多利明. *Keisei hisaku* 経世秘策 [A Secret Plan of Government]. In *Nihon keizai sōsho* 日本経済叢書 [Series on the Japanese Economy] vol. 12, edited by Takimoto Seiichi 瀧本誠一, 49–82. Tokyo: Nihon keizai sōsho kankōkai 日本経済叢書刊行会, 1915.

Ishii Tadayuki nikki 石井忠運日記 [Ishii Tadayuki Diary], written by Ishii Shōzaemon Tadayuki 石井庄左衛門忠運, 18 volumes. Held by Akita-ken kōbunshokan 秋田県公文書館 [Akita Prefecture Archives]. The translation into modern Japanese has been published in *Shin Akita sōsho* 新秋田叢書 [New Akita Series], series 2, vol. 4–8 第 2 期第 4–8 巻. Rekishi toshosha 歴史図書社, 1973–1974.

Isonokami Hachirō [Nobutsugu] 石上八郎 [宣続]. *Bōkaen manroku* 卯花園漫録 [Bōkaen Musings]. In *Shin enseki jisshu* 新燕石十種 5, edited by Kokusho kankōkai 国書刊行会, 154–55. Tokyo: Chūō kōronsha 中央公論社, 1981.

"Junkairoku" 巡海録 [Records of Ocean Travel]. In *Tsūkō Ichiran* 6, 69–70.

Kaban kōeki meisaiki 華蠻交易明細記 [Detailed Record of Trade with China and Europe]. In *Nagasaki kenshi: shiryōhen*, 4 長崎県史 史料編 第四 [Nagasaki Prefectural History: Historical Sources], edited by Nagasaki kenshi hensan iinkai 長崎県史編纂委員会, 245–425. Tokyo: Yoshikawa kōbunkan 吉川弘文館.

Ka-I hentai 華夷変態 [The reversal of *China and* the Barbarians] jō 上 [Vol. 1], edited by Hayashi Harukatsu 林春勝 and Hayashi Nobuatsu 林信篤. Tokyo: Tōyō Bunko 東洋文庫, 1958.

Ka-I hentai 華夷変態 [The reversal of *China and* the Barbarians] ge 下 [Vol. 2], edited by Hayashi Harukatsu 林春勝 and Hayashi Nobuatsu 林信篤. Tokyo: Tōyō Bunko 東洋文庫, 1959.

[Kaiwatashi dōdakachō] *Kanbun sannen udoshi yori nennen tōsen kaiwatashi dōdakachō* 寛文三卯年より年々唐船買渡銅高帳 [Documents on the Annual Quantities of Copper Purchased and Exported by Chinese Ships since 1663]. Sumitomo-ke bunsho 住友家文書 [Sumitomo family documents], 26-6-3-1.

[Kakitsuke] "Saru nedoshi yori tatsudoshi made gokanen-bun kaiire dōdaka narabini kainedan no kakitsuke 去ル子年より辰年まで五ケ年分買入銅高并買値段之書付 [Notes on the comparative sales price of copper in the five-year period between 1708 and 1712]". In *"Dō-kaisho kōyōchō (kyōhō ni-nen)" hoka dō-bōeki kankei shiryō* 「銅会所公用帳(享保二年)」ほか銅貿易関係史料 [Public records from the copper office (1717) and other related material about the copper trade] Sumitomo shiryō sōsho 住友史料叢書 vol. 15, edited by Sumitomo shiryōkan 住友史料館. Kyoto: Shibunkan shuppan 思文閣出版, 2000.

Kinsei fūzoku kenbunshū 近世風俗見聞集 [Collection of Reports on Early Modern Customs] vol. 2, edited by Kokusho kankōkai 国書刊行会. Tokyo: Kokusho kankōkai, 1913.

Kitamura Kōjō 喜多村香城. *Samidare zōshi* 五月雨草紙 [Monsoon Writings]. In *Shin enseki jisshu* 新燕石十種 3, edited by Kokusho kankōkai 国書刊行会: 5–61. Tokyo: Chūō kōronsha 中央公論社, 1981.

Kitagawa Morisada 喜多川守貞. *Ruijū kinsei fūzoku shi* 類従近世風俗志 [Categorized Account of Early Modern Customs]. Tokyo: Uozumi shoten 魚住書店, 1928; repr. 1964.

Kiyō gundan 崎陽群談 [Historical records of Nagasaki and Nagasaki trade as compiled by the Nagasaki-bugyō]. Revised by Nakata Yasunao 中田易直 and Nakamura Tadashi 中村質. Tokyo: Kondō shuppansha 近藤出版社, 1974.

Kokuten ruishō 国典類抄 [Akita-domain's Records: A Compilation Based on Retainers' Diaries and other sources], dai 19 kan 第 19 巻 [vol. 19] zatsubu 雑部二 [various matters 2]. Edited by Akita kenritsu Akita toshokan 秋田県立秋田図書館 [Akita prefecture Akita library], 1984.

Kyōwa sannen idoshichū shuppan hikiawasechō 享和三年亥年中　出帆引合帳 [*Transaction Records of Ship Departures* in the fiscal year of 1803]. Held by the Nagasaki rekishi bunka hakubutsukan 長崎歴史文化博物館 [Nagasaki Museum of History and Culture], ref. Watanabe 316–112 図書番号：渡辺 316–112.

Matsudaira Sadanobu 松平定信. *Uge no hitogoto* 宇下人言 [Memoirs of Matsudaira Sadanobu]. Tokyo: Iwanami shoten 岩波書店, repr. 1970.

Matsura Seizan 松浦静山. *Kasshi yawa* 甲子夜話 [Stories of the Nights of Kasshi] 3, revised by Nakamura Yukihiko 中村幸彦 and Nakano Mitsutoshi 中野三敏, Tōyō Bunko 東洋文庫 321. Tokyo: Heibonsha, 1977.

———. *Kasshi yawa* 甲子夜話 [Stories of the Nights of Kasshi] 4, revised by Nakamura Yukihiko 中村幸彦 and Nakano Mitsutoshi 中野三敏, Tōyō Bunko 東洋文庫333. Tokyo: Heibonsha, 1978.

Miyazaki Yasusada (Antei) 宮崎安貞. *Nōgyō zensho* 農業全書 [Collected Works on Agriculture], vols. 1–5, reprinted in *Nihon nōgyō zenshū* 日本農業全集 12. Tokyo: Shadan hōjin nōsangyoson bunka kyōkai 社団法人農山漁村文化協会, 1978.

Nagasaki hiroku 長崎秘録 [*Secret Records of Nagasaki*] 2. Held by Aichi-ken Kariya-shi toshokan 愛知県刈谷市図書館Record number: to 3.815, satsu 4, mon 4 kōyon, gō 62 図書番号：登3.815、冊4、門 4 甲四、號 62 蔵.

Nagasaki jikki nendairoku 長崎実記年代録 [Epochal Historical Record of Nagasaki], edited by Kyūshū bunkashi kenkyūjo shiryōshū kankōkai 九州文化史研究所史料集刊行会 [Publication of historical sources by the Kyūshū Department of History and Cultural Studies]. Fukuoka: Kyūshū bunka kenkyūjō shiryōshū kankōkai 九州文化史研究所史料集刊行会, 1999.

Nagasaki jitsuroku taisei 長崎実録大成, [Annals of Nagasaki] *seihen* 正編. Nagasaki bunken sōsho vol. 1, no.2 長崎文献叢書第一集第二巻, written by Tanabe Mokei 田辺茂啓, revised by Niwa Kankichi 丹羽漢吉 and Morinaga Taneo 森永種夫. Nagasaki: Nagasaki bunkensha 長崎文献社, 1973.

Nagasaki kaisho gosatsumono 長崎會所五冊物 [*Five-volume record of the Nagasaki Clearing House for Foreign Trade Accounts*]. In: *Nagasaki kenshi: shiryōhen* 長崎県史　資料編　第四 [Nagasaki prefectural history: Historical sources] vol. 4, edited by Nagasaki kenshi hensan iinkai 長崎県史編纂委員会, 1–243. Tokyo: Yoshikawa kōbunkan 吉川弘文館, 1965.

Nagasaki Oranda shōkan no nikki 長崎オランダ商館の日記 [Diary of the Dutch Trading House at Nagasaki] vol. 1, edited by Nederlandsche Oost-Indische Compagnie Comptoir Nangasackij, translated by Murakami Naojirō 村上直次郎. Tokyo: Iwanami shoten 岩波書店, 1956; repr. 1980.

Nagasaki Oranda shōkan no nikki 長崎オランダ商館の日記 [Diary of the Dutch Trading House at Nagasaki] vol. 2, edited by Nederlandsche Oost-Indische Compagnie Comptoir Nangasackij, translated by Murakami Naojirō 村上直次郎. Tokyo: Iwanami shoten 岩波書店, 1957; repr. 1980.

Nagasaki Oranda shōkan no nikki 長崎オランダ商館の日記 [Diary of the Dutch trading house at Nagasaki] vol. 3, edited by Nederlandsche Oost-Indische Compagnie Comptoir Nangasackij, translated by Murakami Naojirō 村上直次郎. Tokyo: Iwanami shoten 岩波書店, 1958; repr. 1980.

Nakai Chikuzan 中井竹山. *Sōbōkigen* 草芽危言 [Civilian Critiques on National Politics]. In *Nihon keizai sōsho* 日本経済叢書 [Series on the Japanese Economy] vol. 16, edited by Takimoto Seiichi 瀧本誠一, 259–490. Tokyo: Nihon keizai sōsho kankōkai 日本経済叢書刊行会, 1915.

Ofuregaki kampō shūsei 御触書寛保集成 [Ordinances (of the Tokugawa shogunate) compiled in the Kampō Period (1741–1743)], edited by Takayanagi Shinzō 高柳真三 and Ishii Ryōsuke 石井良助. Tokyo: Iwanami shoten 岩波書店, 1958.

Okamoto Matatarō Mototaka nikki 岡本又太郎元貴日記 [Diary of Okamoto Matatarō Mototaka]. In *Kokuten ruishō*, zatsubu 雑部二 [various matters 2], 492–493.

Ōkura Nagatsune 大蔵永常. *Kōeki kokusan kō* 広益国産考 [On Increasing Profits and National Production], reprinted in *Nihon nōgyō zenshū* 日本農業全集 14. Tokyo: Shadan hōjin nōsangyoson bunka kyōkai 社団法人農山漁村文化協会, 1978.

Oyamada Tomokiyo 小山田與清. *Matsuya hikki* 松屋筆記 [Matsuya Notes] 2, edited by Ichijima Kenkichi 市島謙吉. Tokyo: Kokusho kankōkai 国書刊行会, 1908.

Satake kafu 佐竹家譜 [Genealogical Tree of Family Satake], revised by Hara Takeo 原武男, vol. 2. Tōyō shoin 東洋書院, 1989.

Shangyu Tiaoli 上諭條例 [Edicts and Regulations]. Printed and published by the Qing Dynasty Jiangsu administrative commissioner's office] 清刊（江蘇布政使司), 1735–1750. Held by the National Archives of Japan 国立公文書館内閣文庫蔵, No. shō 64-287-19 図書番号：昌 64-287-19.

Shinkoku Tenshin shijō kaisanbutsu keikyō 清國天津市場海産物景況 (明治 19 年（光緒十二 1886）8月28日付の在天津帝國領事館報告 [A Report from the Japanese Imperial Consulate in Tianjin Dated 28 August 1886]. In *Tsūshōisan* 通商彙纂. Tokyo: Gaimushō tsūshōkyoku 外務省通商局, 1886).

Shōdai sōsho zokuhen 昭代叢書續編. See *Zhaodai congshu xubian* in 1.2 Chinese primary sources.

[Shokoku dōzan] "Shokoku dōzan shutsudō-daka no oboe 諸国銅山出銅高の覚 [Record of Provincial Mine Output Levels]". In *Dōza kōyōdome, Dōza goyō hikae* 銅座公用留; 銅座御用扣 [Public records and transcripts of the Copper Agency] Sumitomo shiryō sōsho 住友史料叢書 vol. 4, edited by Sumitomo shiryōkan 住友史料館. Kyoto: Shibunkan shuppan 思文閣出版, 1989.

Shōsō sanbō shishū 小倉山房詩集. See *Xiaocang shanfang shiju* in 1.2 Chinese primary sources.

Shōsō sanbō shishū, hoi 小倉山房詩集， 補遺. See *Xiaocang shangfang shiji buyi* in 1.2 Chinese primary sources.

Suijinroku 吹塵錄 [Records of Blowing Away Dust]. In *Katsu Kaishū zenshū* 勝海舟全集, vol. 9. Tokyo: Keisō shobō, 1976.

Takizawa Bakin 滝沢馬琴, *Ibun zakkō* 異聞雑稿 jō 上冊 [Miscellanous Drafts of Strange Tales, vol. 1]. In *Zoku enseki jisshu* 続燕石十種 2, edited by Iwamoto Kattōshi 岩本活東子, 209–54. Tokyo: Chūō kōronsha 中央公論社, 1980.

Tōban kamotsu chō 唐蠻貨物帳 [Lists of Cargo from China and Europe], Parts 1 and 2 上下. Tokyo: Naikaku Bunko 内閣文庫, 1970.

Tokugawa jikki 徳川実記 [The true Tokugawa records] vol. 9. In Kuroita Katsumi 黒板勝美 ed., *Kokushi taikei* 国史大系 46. Tokyo: Yoshikawa kōbunkan 吉川弘文館, 1982.

Tōsen shinkō kaitōroku 唐船進港回棹録; *Shimabarabon Tōjin fūsetsugaki* 島原本唐人風説書; *wappu tomechō* 割符留帳 [Record of Arrivals and Departures of Chinese Ships; Tales of China, Shimabara Volume; Record of Trade Permits], edited by Ōba Osamu 大庭脩. Osaka: Kansai Daigaku tōzai gakujutsu kenkyūjo 関西大学東西学術研究所 [Kansai University Press], 1974.

Tsūko ichiran 通航一覧 [Collected Diplomatic and Trade Records] 5, edited by Hayashi Akira 林煌. Tokyo: Kokusho kankōkai 国書刊行会, 1913.

Tsūko ichiran 通航一覧 [Collected Diplomatic and Trade Records] 6, edited by Hayashi Akira 林煌. Tokyo: Kokusho kankōkai 国書刊行会, 1913.

Tō tsūji kaisho nichiroku 唐通事會所日録 [*Daily records of the center of Chinese interpreters*]. Edited by Tokyo Daigaku Shiryō Hensansho 東京大学史料編纂所. Tokyo: Tokyo Daigaku Shuppankai 東京大学出版会, 1955–68; repr. 1984.

Tsuchiya Tomotora nikki 土屋知虎日記 [Tsuchiya Tomotora Diary], written by Tsuchiya Tomotora 土屋知虎. Held by Akita-ken kōbunshokan 秋田県公文書館 [Akita Prefecture Archives].

Umezu Tadayoshi nikki 梅津忠宴日記 [Umezu Tadayoshi Diary], written by Umezu Tadayoshi 梅津忠宴. Held by Akita-ken kōbunshokan 秋田県公文書館 [Akita Prefecture Archives].

Wakan sansai zue 和漢三才図会 [Illustrated Sino-Japanese Encyclopedia] 16, edited by Shimada Isao 島田勇雄, Takeshima Atsuo 竹島淳夫 and Higuchi Motomi 樋口元巳, Tōyō Bunko vol. 521. Tokyo: Heibonsha 平凡社, 1990.

Wakan sansai zue 和漢三才図会 [Illustrated Sino-Japanese Encyclopedia] 18, edited by Shimada Isao 島田勇雄, Takeshima Atsuo 竹島淳夫 and Higuchi Motomi 樋口元巳, Tōyō Bunko vol. 532. Tokyo: Heibonsha 平凡社, 1991.

"Zokudankai" 續談海 [Sea of Historical Stories—Sequel]. In *Tsūkō Ichiran*, 6: 94.

Zoku Nagasaki jitsuroku taisei 続長崎実録大成 [Sequel to Annals of Nagasaki]. Nagasaki bunken sōsho vol. 1, no. 4 長崎文献叢書第一集第四巻, edited by Kohara Hazan 小原克紹. Nagasaki: Nagasaki bunkensha 長崎文献社, 1974.

1.2 *Chinese Primary Sources*

Biji xiaoshuo daguan 筆記小説大観 [*Complete Collection of Literary Jottings and Informal Stories*] vol. 27. Reprint, Yangzhou: Jiangsu Guangling guji keyin chubanshe 江蘇廣陵古籍刻印出版社, 1984.

Cai Yurong 蔡毓榮, "Chou Dian shi shu" 籌滇十疏 [Ten Documents for Governing Yunnan], document 4: "I Licai" [Finance]. In *Yunnan tongzhi* 雲南通志 (Provincial Gazetteers of Yunnan) 1736 ed., 30 vols. Reprint, edited by E. Ertai 鄂爾泰 et al. vol. 29, 17–23. Jiangsu: Guangling guji keyin she 江蘇廣陵古籍刻印社, 1988.

Chenghai-xianzhi 澄海縣志 [Chenghai County Gazetteer], vol. 6, edited by Li Shuji 李書吉 et al. Reprint, Taiwan 台湾: Chengwen chubanshe 成文出版社, 1967.

Cheng Jiatang shizai daomai gongju tongjin jielüe 程稼堂實在盜賣公局銅觔節略 [Description of the Illegal Sale of Quota Copper by Cheng Jiatang]. Manuscript, 1861, held by the Nagasaki Museum of History and Culture (Nagasaki rekishi bunka hakubutsukan 長崎歴史文化博物館).

Fujian shengli 福建省例 [Laws and Regulations of Fujian Province]. Taiwan Wenxian Congkan, no. 199. Taipei: Taiwan yinhang jingji yanjiushi, 1964.

"Gokaen zuihitsu" 護花園随筆 [Gokaen Essays]. In *Tsūkō ichiran* 6, 56.

[GDKZ] *Gongzhong dang Kangxichao zouzhe* 宮中檔康熙朝奏摺 [Secret Palace Memorials of the Kangxi Period]. Taipei: Guoli gugong bowuyuan 故宮博物院 [National Palace Museum], 1976–77.

[GDYZ] *Gongzhong dang Yongzhengchao zouzhe* 宮中檔雍正朝奏摺 [Secret Palace Memorials of the Yongzheng Period]. Taipei: Guoli gugong bowuyuan 故宮博物院 [National Palace Museum], 1977–80.

[GDQZ] *Gongzhong dang Qianlongchao zouzhe* 宮中檔乾隆朝奏摺 [Secret Palace Memorials of the Qianlong Period]. Taipei: Guoli gugong bowuyuan 故宮博物院 [National Palace Museum], 1982–88.

[CJD] *Guanyu Jiangning zhizao Cao jia dangan shiliao* 關於江寧織造曹家檔案史料 [Archival Materials relating to the Nanjing Superintendants of Imperial Silk Manufacturing Cao Yin and his Sons]. Beijing: Zhonghua shuju, 1975.

Gongzhongdang Qianlongchao zouzhe 宮中档乾隆朝奏摺 [*Palace Memorials of the Qianlong Reign*]. Taipei: Guoli gugong bowuyuan 故宮博物院 [Taipei National Palace Museum], 1972.

Huangchao jingshi wenbian 皇朝經世文編 [Collected Writings on Statecraft during the Qing Dynasty], edited by He Changling 賀長齡 et al., 1827ed, 120 vols. Reprint, Beijing: Zhonghua shuju, 1992.

Huangchao zhanggu huibian 皇朝掌故彙編 [Collected Historical Facts and Records of the Qing Dynasty]. 1902 ed. Reprint, Taipei: Wenhai chubanshe, 1964.

Kangxi qijuzhu 康熙起居注 [Daily Activities of the Kangxi Emperor], edited by First Historical Archives 中國第一歷史檔案館. Beijing: Zhonghua shuju, 1984.

Laoxuean biji 老學庵筆記 [Jottings by Lu You], edited by Li Jianxiong *李劍雄 and Liu Dequan 劉德權*. Tang Song shiliao biji congkan 唐宋史料筆記叢刊. Beijng: Zhonghua shuju 中華書局, 1979.

Lidai baoan 歷代寶案 [The Precious Documents of Successive Dynasties of the Ryūkyū Kingdom]. Taipei: National Taiwan University, 1972.

Lin Wenzhonggong zhengshu jiaji 林文忠公政書 甲集 [Political Correspondence of Lin Tse-hsu]. Reprint, Taipei: Wenhai chubanshe, 1967.

[MQD] *Ming-Qing dangan* 明清檔案 [Archives from the Ming-Qing], edited by Zhang Weiren 張偉仁, 324 vols. Taipei: Institute of History and Philology, Academia Sinica, 1986–1995.

Qingchao xu wenxian tongkao 清朝續文獻通考 [Institution, Finance and Organisation of the Late Qing Dynasty] vol. 3 (4 volumes in total), written by Shitong 十通. Hangzhou 杭州: Zhejiang Guji Chuban-she 浙江古籍出版社, 2000.

Qianlong chao shangyu dang 乾隆朝上諭檔 [Collected Imperial Orders], comp. by First Historical Archives. Reprint, Beijing: Dangan chubanshe, 1991.

Qinding Da Qing huidian 欽定大清會典 [Imperially Endorsed, Collected Administrative Statutes of the Great Qing], 1818 ed. Reprint, Taipei: Wenhai chubanshe, 1991.

Qinding Da Qing huidian shili 欽定大清會典事例 [Imperially Endorsed, Collected Administrative Statutes and Precedents of the Great Qing], 1899 ed. Reprint, Beijing: Zhonghua shuju, 1991.

Qinding Da Qing huidian zeli 欽定大清會典則例 [Imperially Endorsed, Collected Administrative Statutes and Precedents of the Great Qing], photocopy of Weng Yuan Ge, Complete Collection in Four Treasuries, vol. 621. Taipei: Shangwu yinshuguan, 1983.

Qianlong chaocheng bantan wu dangan zhuanbian 乾隆朝懲辦貪污檔案撰編 [Official Documents on the Corruption of Public Servants during the Qinglong Period (1736–1795)] vol. 2 第二冊, edited by Zhongguo-diyi-lishi-danganguan 中国第一歴史檔案館. Beijing: Zhonghua shuju 中華書局, 1994.

Qing-shi-gao 清史稿 [History of the Qing Dynasty], comp. by Zhao Erxun 趙爾巽 et al., vol. 36. Reprint, Beijing: Zhonguha shuju 中華書局, 1977.

[QWT] *Qingchao wenxian tongkao* 清朝文獻通考 [Encyclopedia of the Historical Records of the Qing Dynasty, also translated as Comprehensive Scrutiny of Documents: Qing Dynasty]. Reprint, Taipei: Xinxing shuju, 1963.

Qingshi liezhuan 清史列傳 [Qing Historical Biographies]. Beijing: Zhonghua shuju, 1987.

Qing shilu 清實錄 [Veritable Records of the Great Qing]. Kangxi and Qianlong period. Reprint, Beijing: Zhonghua shuju, 1986.

Rekidai hōan 歴代寶案 [*Precious Documents of Successive Generations*]. Edited by Okinawa prefectural Library, Shiryō henshūshitsu 沖縄県立図書館史料編集室, revised edition, vol. 4. Naha: 沖縄県教育委員会 Okinawa ken kyōiku iinkai, 1993.

Riben-fengtu-ji 日本風土記 (*Nippon fudoki*) [Description of Regional Climate and Culture of Japan] vol. 1, in the supplement to the *Quanzhe-bingzhi-kao* 全浙兵制考 (*zensetsu heiseikō*) [About coastal defence along the Zhejiang coast during the Ming Dynasty], written at the end of the Ming Dynasty: page unknown. Held by National Archives of Japan, Cabinet Library 国立公文書館内閣文庫.

Suzhou fuzhi 蘇州府志 [Suzhou Prefectural Gazetteer] 1883 ed. Reprint, Taipei: Chengwen chubanshe, 1970.

Xiaocang shanfang shiju 小倉山房詩集 [Collected *Xiaocang-shan-fang* Poems]. Library identification number *zhang* (長) 37124–33, dimensions 16.2 cm wide × 24.9 cm high, ten volumes, held by Shanghai Library 上海図書館.

Xiaocang shangfang shiji buyi 小倉山房詩集, 補遺 [Collected *Xiaocang-shan-fang* Poems, Supplement]. Library identification number *zhang* (長) 334533–36, dimensions 11.8 cm wide × 17.5 cm high, four volumes, small-sized version, held by Shanghai Library 上海図書館.

Wang Taiyue, 王太岳. "Tongzheng yi" 銅政議 [Deliberations on the Copper Administration]. In *Huangchao jingshi wenbian* 皇朝經世文編 [Collected Writings on Statecraft during the Qing Dynasty] edited by He Changling 賀長齡 et al. 1827 ed., 120 vols., vol. 52, 17–34. Reprint, Beijing: Zhonghua shuju, 1992.

Wang Xiqi 王錫祺 (ed.). *Xiaofanghuzhai yudi congchao* 小方壺齋輿地叢鈔 [*Geographical Essays from the Xiaofanghu Studio*]. Taipei: Guangwen shuju 廣文書局, 1962 (Copy book 影印本).

Yan Zhongping 嚴中平. *Qingdai Yunnan tongzheng kao* 清代雲南銅政考 [*An Investigation of the Copper Administration of Qing-era Yunnan*]. Peking 北京: Zhonghua shu ju 中華書局, 1948.

Zhaodai congshu xubian 昭代叢書續編 [Collection of Poems, Diaries, Records, Stories about Homecoming and Essays during the Qing Dynasty], issue 219, written during the Daoguang Era (1821–1850), held by Kansai University Library, Hakuen Library 関西大学図書館泊園文庫.

Zhongguo difangzhi jicheng xiangzhenzhi zhuanji 中国地方志集成　郷鎮志専輯 20 [A Comprehensive Collection of Chinese Local Gazetteers: Local Gazetteers of Villages and Market Towns, vol. 20]. Edited by Lu Jianxing 盧建幸, Wang Haizhi 王海治 et al. Shanghai: Jiangsu Guji Chubanshe 江蘇古籍出版社, Shanghai Shudian 上海書店, Bashu Shushe 巴蜀書社, 1992.

Zhupi zouzhe caizheng lei mulu 硃批奏摺財政類目錄 [Imperially Endorsed Memorials, Financial Matters, Table of Contents]. First Historical Archives (comp.), vol. 4. Beijing: Zhongguo caizheng jingji chubanshe, 1992.

[ZZCL] *Zhupi zouzhe caizheng lei* 硃批奏摺財政類 [Imperially Endorsed Memorials, Financial Matters], microfilm, held by First Historical Archives, Beijing.

2 Secondary Sources

Akitaken 秋田県 (ed.). *Akitaken shi shiryōhen* 秋田県史資料編 [History of Akita Prefecture: Sources], *vol. 3 kinsei ge* 近世下 [Early Modern Sources], edited by Akitaken 秋田県. Tokyo: Akitaken, 1963.

———. *Akitaken-shi* 秋田県史 (History of Akita Prefecture) vol. 3. Tokyo: Akitaken, 1965.

Arai Eiji 荒居英次. *Kinsei kaisanbutsu bōeki shi no kenkyū—Chūgokumuke yushutsu bōeki to kaisanbutsu* 近世海産物貿易史の研究一中国向け輸出貿易と海産物—[Studies on the History of the Trade in Marine Products in the Early Modern Period: Marine Products and the Export Trade to China]. Tokyo: Yoshikawa kōbunkan 吉川弘文館, 1975.

Arano Yasunori 荒野泰典. "18seiki no Higashi-Ajia to Nippon" 十八世紀の東アジアと日本 [East Asia and Japan during the 18th century]. In Rekishi-gaku kenkyūkai

歴史学研究会 and Nihons-shi kenkyūkai 日本史研究会 (eds.) *Kōza Nihon rekishi 6 kinsei 2* 講座日本歴史 6 近世 2 [Series on Japanese History vol. 6, Early mordern period no. 2], 1–44. Tokyo: Tokyo daigaku shuppan 東京大学出版 [University Press], 1984.

——— (ed.). *Sakoku no kenkyū* 鎖国の研究 [Research on so-called seclusion period]. Tokyo: Popurasha ポプラ社, 2008.

Asakura Haruhiko 朝倉治彦. "Kōki 後記 [Afterword]". In *Shin enseki jisshu* 新燕石十種 3, edited by Kokusho kankōkai 国書刊行会, 335–359. Tokyo: Chūō kōronsha 中央公論社, 1981.

———. "Kōki 後記 [Afterword]". In *Shin enseki jisshu* 新燕石十種 5, edited by Kokusho kankōkai 国書刊行会, 407–11. Tokyo: Chūō kōronsha 中央公論社, 1981.

———. "Kōki 後記 [Afterword]". In "Afterword," In *Zoku enseki jisshu* 続燕石十種 2, edited by Iwamoto Kattōshi 岩本活東子, 427–36. Tokyo: Chūō kōronsha 中央公論社, 1980.

Chang Ling 常玲. "Qingdai Yunnan de 'fangben shoutong' zhengce" 清代雲南的"放本收銅"政策 [The Policy of Providing Capital to Commission Businessmen to Mine Copper for the State in Qing Era Yunnan]. *Sixiang zhanxian* 思想戰線 [Thought Frontline] 2 (1988): 85–89.

Chen Ciyu 陳慈玉. "18 Shiji Zhongguo Yunnan de tong shengchan" 18 世紀中國雲南的銅生產 [Yunnan Copper Production in 18th Century China]. In *Guoshi shilun* 國史釋論 [Collection of Essays on Chinese History], vol. 1, edited by Yang Liansheng 楊聯陞 et al., 283–99. Taipei: Shihuo chubanshe, 1987.

Chen Xude 陳序德. "Qingchu Jinshajiang shuidao kaijun gaikuang" 清初金沙江水道開浚概況 [The Opening of the Jinsha River Water Route in the Early Qing Dynasty]. In *Zaotong wenshi ziliao xuanji* 昭通文史資料選輯 [Selected Data on the History of Zhaotong County] 3, edited by Zhongguo renmin zhengzhi xieshang huiyi Yunnansheng Zaotongshi weiyuanhui wenshi ziliao bianjishi 中國人民政治協商會議雲南省昭通市委員會文史資料編輯室, 200–5. Zaotong: Zhongguo renmin zhengzhi xieshang huiyi Yunnansheng Zaotongshi weiyuanhui wenshi ziliao bianjishi, 1988.

Cheng Wingcheong 鄭永昌. *Mingmo Qingchu de yingui qianjian xianxiang yu xiangguan zhengzhi jingji sixiang* 明末清初的銀貴錢賤現象與相關政治經濟思想 [The Seventeeth Century Monetary Crisis and the Related Political-Economic Ideas in China], Monograph Series no. 24. Taipei: Institute of History of National Taiwan Normal University, 1994.

Chiu Pengsheng 邱澎生. "Shiba shiji Yunnan tongcai shichangzhong de guanshang guanxi yu liyi guannian" 十八世紀雲南銅材市場中的官商關係與利益觀念 ['Interests' in Economic Organization: The Shaping of the Yunnan Copper Market in Eighteenth-Century China]. *Zhongyang yanjiuyuan lishi yuyan yanjiusuo jikan* 中央

研究院歷史語言研究所集刊 [Bulletin of the Institute of History and Philology Academia Sinica], 72–1 (2001): 49–121.

Ch'üan Hansheng 全漢昇. "Qingdai Yunnan tongkuang gongye" 清代雲南銅礦工業 [The Yunnan Copper Mining Industry in the Qing Era]. *Zhongguo wenhua yanjiusuo xuebao* 中國文化研究所學報 [Journal of the Institute of Chinese Culture] of the Chinese University of Hong Kong, 7, No. 1 (1974): 155–82.

Dunstan, Helen. "Safely Supping with the Devil: The Qing State and Its Merchant Suppliers of Copper." *Late Imperial China* 13: 2 (1992): 42–81.

Flynn, Dennis O. and Arturo Giráldez. "Cycles of Silver: Global Economic Unity through the Mid-Eighteenth Century." *Journal of world history* 13, 2 (2002): 391–427.

Gokyū Sessō 五弓雪窓, ed. *Jijitsu bunpen* 事実文編 [Collection of biographies of Japanese from the 16th century until 1881] vol. 3. Osaka: Kansai Daigaku Shuppan Kōhōbu 関西大学出版広報部, 1980.

Glahn, Richard von. *Fountain of Fortune. Money and monetary policy in China, 1000–1700*. Berkeley, Los Angeles, London: University of California Press, 1996.

Gramlich-Oka, Bettina. "Shogunal Administration of Copper in the Mid-Tokugawa Period (1670–1720)." In *Metals, Monies, and Markets in Early Modern Societies: East Asian and Global Perspectives*, edited by Thomas Hirzel and Nanny Kim: 65–105. Berlin: LIT Verlag, 2008.

Hall, John W. "Notes on the Early Ch'ing Copper Trade with Japan." *Harvard Journal of Asiatic Studies*, 12 (1949): 444–61.

Huang Xiaoping 黃小平. "E Ertai yu Yun-Gui Miaojiang de jingying, 1726–1731" 鄂爾泰與雲貴苗疆的經營 [E Ertai and the Management of the Miao Border Area in Yunnan and Guizhou]. *Dalu zazhi* 大陸雜誌 [Mainland China Magazine] 55: 3 (1977): 19–45.

Ichijima Kenkichi 市島謙吉. "Reigen 例言 [Preface]". In *Matsuya hikki* 松屋筆記 1, edited by Ichijima Kenkichi 1: 1–5. Tokyo: Kokusho kankōkai 国書刊行会, 1908.

Ichikawa San'yō 市河三陽, ed. *Kansai yokō* 寛斎餘稿 [Collection of writings of Ichikawa Kansai]. Tokyo: Yūtokuen 遊徳園, 1926.

Imai Noriko 今井典子. "Kinsei chūki no chiuridō ni tsuite" 近世中期の地売銅について [About copper for domestic consumption in the middle of the early mordern period]. *Nihon rekishi* 日本歴史 5 (May 1988): 50–67.

———. "Kinsei Osaka no dō-kanren-gyōsha" 近世大坂の銅関連業者 [Copper Suppliers of early modern Osaka]. In *Osaka-shi bunkazai kyōkai kenkyū kiyō* 2 (1999): 193–202.

———. "Hōei, Shōtoku-ki no dōbōeki to Sumitomo 宝永・正徳期の銅貿易と住友 [Sumitomo and the copper trade in the Hōei and Shōtoku period]". *Sumitomo shiryōkanpō* 住友史料館報 No. 33 (2002): 1–38.

———. "Nagasaki bōeki taisei to Genbun dōza 長崎貿易体制と元文銅座 [Nagasaki trade system and Genbun Copper Agency]." *Sumitomo shiryōkanpō* 住友史料館報 No. 38 (2007): 59–84.

———. "Kinsei Nihon no dō: dōshijō to dōtōsei" 近世日本の銅ー銅市場と銅統制—[Copper in Early Modern Japan: Copper Market and Copper Monopoly]. *Sumitomo shiryōkanpō* 住友史料館報 [Journal of Sumitomo Archives] 39 (2008): 13–65.

Innes, Robert LeRoy. *The Door Ajar: Japan's Foreign Trade in the Seventeenth Century* (Authorized facsimile made from the microfilm master copy of the original Phil. D. Thesis published by UMI. Printed in 2006). Ann Arbor, Michigan: UMI Dissertation Services, 1980.

Ishii Takashi 石井孝. *Bakumatsu kaikōki keizeishi kenkyū* 幕末開港期経済史研究 [Economic historical study of the opening of Japan and the fall of the shogunate] Tokyo: Yūrindō 有隣堂, 1987.

———. "Bakumatsu ni okeru bakufu no dō yushutsu kinshi seisaku" 幕末における幕府の銅輸出禁止政策 [The Policy of Banning Copper Exports in the Bakufu Period]. *Rekishigaku kenkyū* 歴史學研究 [Studies in History] 130 (1947): 1–32.

Ishikawa Junkichi 石川準吉. *Ikuno ginzan to ikuno daikan* 生野銀山と生野代官 [The Silver Mines of Ikuno and the Ikuno Magistrate]. Tokyo: Nihon kōgyō shinbunsha 日本工業新聞社, 1974.

Iwao Seiichi 岩生成一. "Edō jidai ni okeru dōsen no kaigai yushutsu ni tsuite" 江戸時代における銅銭の海外輸出について [On the export of copper coins in the Tokugawa period]. *Shigaku zasshi* 39 (1928): 99–110.

———. "Kinsei Nisshi bōeki ni kansuru sūryō teki kōsatsu" 近世日支貿易に關する數量的考察 [A study on the Chinese trade with Japan in the 17th century, with focus on volume and quantity] *Shigaku zasshi* 史學雜誌 [Journal of History] 62: 11 (Nov. 1953): 981–1020.

———. "Edo jidai no satō bōeki ni tsuite" 江戸時代の砂糖貿易について [The Edo-period sugar trade]. *Nihon Gakushiin kiyō* 日本学士院紀要 31, 1 (March 1973), 1–33.

Iwasaki Yoshinori 岩﨑義則. *Kinsei dōza no kenkyū* 近世銅座の研究 [Research on the Copper Agency in the early modern period]. PhD diss., Kyūshū University, 1997.

———. "Nagasaki kaidō to sono kikō—kaisen chōtatsu to Nagasaki kaidō—" 長崎廻銅とその機構ー廻船調達と長崎廻銅—[Copper transport to Nagasaki and its structure—procurement of vessels and copper transport]. In *Sakoku to kokusai kankei* 鎖国と国際関係 [National Seclusion and International relationships], edited by Nakamura Tadashi 中村質, 356–376. Tokyo: Yoshikawa kōbunkan 吉川弘文館, 1997.

———. "Kampō Kan'ei-ki no bōeki kaikaku to dō tōseisaku 寛保-寛延期の貿易改革と銅統制策 [The copper control policies and the trade reforms of the Kampō and Kanei period]". *Sumitomo shiryōkanpō* 住友史料館報 No. 31 (2000): 39–68.

———. "Genbun dōza-ki no Nagasaki kaidō to Nagasaki bōeki" 元文銅座期の長崎廻銅と長崎貿易 [Copper transport to Nagasaki and trade in Nagasaki during the period of the Genbun Copper Office]. *Sumitomo shiryōkanpō* 住友史料館報 34 (2003): 1–42.

———. "Kinzoku-shi kinsei Nihon-dō no seisan to ryūtsū" 金属史　近世日本銅の生産と流通 [A History of Metal: Production and Distribution of Copper in Early Modern Japan]. Nihon-shi no kenkyū 日本史の研究 235 (December 2011): 29–43.

Kawakatsu Mamoru 川勝守. "Shin Kenryūki Unnandō no keiun mondai" 清乾隆期雲南銅の京運問題 [The Problem of Shipment of Yunnan Copper to Beijing in the Qianlong Reign of the Qing Dynasty] *Kyūshū daigaku tōyōshi ronshū* 九州大学東洋史論集 [Kyushu University Collected Essays on Oriental History], 17 (1989): 1–94.

———. "Shin Kenryū shonen Unnandō no Chōkō yuso to toshi Kanko" 清乾隆初年雲南銅の長江輸送と都市漢口 [The Transport of Yunnan Copper on the Yangzi River and the Development of Hankou City in the Early Qing Dynasty]. In *Higashi Ajia ni okeru seisan to ryūtsū no rekishi shakaigaku teki kenkyū* 東アジアにおける生產と流通の歴史社會學的研究 [Studies in the Social History of Production and Transportation in East Asia], edited by Kawakatsu Mamoru 川勝守, 406–28. Fukuoka: Chūgoku shoten, 1994.

Kawashima Motojirō 川島元次郎. "Saisho ni kokoromita Shanhai bōeki" 最初に試みた上海貿易 [First Attempts at Trade with Shanghai]. In *Nankoku shiwa* 南國史話 [*Historical Tales from the Southern Lands*], edited by Kawashima Motojirō, 115–66. Tokyo: Heibonsha 平凡社, 1926.

Kim, Nanny. "*Copper Transports* out of Yunnan, ca. 1750–1850: *Transport* Technologies, Natural Difficulties and Environmental Change in a Southwestern Highland Area." In *Metals, Monies, and Markets in Early Modern Societies: East Asian and Global Perspectives*. Monies, Markets and Finance in China and East Asia, vol. 1, edited by Thomas Hirzel and Nanny Kim, 191–220. Berlin: LIT Verlag, 2008.

Kobata Atsushi 小葉田淳. "Satō no shiteki kenkyū ni tsuite" 砂糖の史的研究に就いて Research on the History of Sugar]. In Kobata Atsushi, *Shisetsu Nihon to minami shina* 史説日本と南支那 [History: Japan and Southern China], 213–244. Taipei: Noda shobō 野田書房, 1942.

———. *Kōzan no rekishi* 鉱山の歴史 [History of Mining]. First edition in 1956. Tokyo: Shibundō 至文堂, 1962.

———. *Nihon kōzanshi no kenkyū* 日本鉱山史の研究 [Research in the History of Mining in Japan]. Tokyo: Iwanami shoten 岩波書店, 1968.

———. *Nihon dōkōgyōshi no kenkyū* 日本銅鉱業史の研究 [Research in the History of Copper Mining in Japan]. Kyoto: Shibunkaku shuppan 思文閣出版, 1993.

———. "17seiki narabini 18seiki shoki ni okeru Chūgoku-sen, Oranda-sen ni taisuru wagakuni no dō-yushutsu 17 世紀並びに 18 世紀初期における中国船・オラ

ンダ船に対するわが国の銅輸出 [The Export of Japanese Copper on Chinese and Dutch Ships during the Seventeenth and Early Eighteenth Centuries]. *Nihon kōgyōshi kenkyū* 日本鉱業史研究 No. 31 (1996): 3–6.

Kokushi daijiten henshū iinkai 国史大辞典編集委員会, ed. *Kokushi daijiten* 国史大辞典 [Great Dictionary of National History]. Tokyo: Yoshikawa kōbunkan 吉川弘文館, 1979–1997.

Kōsaka Masanori 香坂昌紀. "Shindai zenki no kansa bendōsei oyobi shōnin bendōsei ni tsuite" 清代前期の關差辦銅制及び商人辦銅制について [On the Systems of Official and Merchant Copper Procurement in the Early Qing]. *Tōhoku gakuin daigaku ronshū rekishigaku chirigaku* 東北学院大学論集 歴史学地理学 [Tohoku Gakuin University Studies in History and Geography], 11 (1981): 115–53.

Kudō Mutsuo 工藤睦男. "Nanbu-han goyōdō-sen Tokujumaru no nansen shimatsuki" 南部藩御用銅船徳寿丸の難船始末記 [Account of the Shipwreck of the Official Copper Carrier Tokujumaru]. *Hirosaki daigaku kyōiku-gakubu kiyō* 弘前大学教育学部紀要 29 (1973): 81–92.

Kudō Yoshishirō 工藤由四郎. *Ani-ai kyōdoshi* 阿仁合郷土誌 [History of Ani-ai]. Akita: Kudō Yoshishirō, 1962.

Lan Yong 藍勇. "Qingdai Diantong jingyong luxian kaoshi" 清代滇銅京運路線考釋 [An Investigation of the Copper Transport Route from Yunnan to Beijing]. *Lishi yanjiu* 歷史研究 [Historical Studies] 3 (2006): 86–97.

Liu Shiuhfeng 劉序楓. "Shinnichi bōeki no yōdōshō ni tsuite: Kenryū-Kanbō ki no kanshō, minshō o chūshin ni" 清日貿易の洋銅商について： 乾隆～咸豐期の官商．民商を中心に [Copper Traders in Sino-Japanese Trade in the Qing Dynasty: A Focus on Official and Private Merchants in the Qianlong and Xianfeng Periods]. *Kyūshū daigaku tōyōshi ronshū* 九州大学東洋史論集 [Kyushu University Essays on Oriental History], 15 (1986): 107–52.

———. "Shindai zenki no Fukken shōnin to Nagasaki bōeki" 清代前期の福建商人と長崎貿易 [Trade between Nagasaki and Fujian Merchants in the Early Qing]. *Kyushū daigaku tōyōshi ronshū* 九州大学東洋史論集 [Kyushu University Essays on Oriental History] 16 (1988): 133–61.

———. "Jūshichi jūhachi-seiki no Chūgoku to Higashi Ajia: Shinchō no kaigai bōeki seisaku o chūshin ni" 17・8世紀の中國と東アジア： 清朝の海外貿易政策を中心に [China and East Asia in the 17th and 18th Centuries: A Focus on the External Trade Policy of the Qing Dynasty]. In *Ajia kara kangaeru 2 Chiiki shisutemu* アジアから考える2 地域システム [The Asian Regional System, #2 in the Asian Perspectives Series], edited by Hamashita Takeshi 濱下武志 et al., 87–129. Tokyo: Tokyo University Press, 1993.

———. "Qingdai de Zhapugang yu Zhong-Ri maoyi" 清代的乍浦港與中日貿易 [Qing Era Zhapu Port and Sino-Japanese Trade] In *Zhongguo haiyang fazhanshi lunwenji* 中國海洋發展史論文集 [Essays on Chinese Maritime History], vol. 5, edited by Chang Pintsun 張彬村 and Liu Shihchi 劉石吉, 187–244. Taipei: Sun Yat-sen Institute for Social Science and Philosophy, Academia Sinica, 1993.

———. "Kinsei Nisshin bōeki to sono kessai" 近世日清貿易とその決済 [Procedures for Settling Accounts for Commerce between Qing-Era China and Japan] *Museum Kyūshū* (Fukuoka: The Kyushu Association for the Promotion of the Establishment of Museums) 51 (Jan 1996): 28–38.

———. "Kyōhō nenkan no Tōsen bōeki to Nihondō" 享保年間の唐船貿易と日本銅 [The Chinese Junk Trade and Japanese Copper in the Kyōhō Period]. In *Sakoku to kokusai kankei* 鎖国と国際関係 [National Seclusion and International Trade], edited by Nakamura Tadashi 中村質, 296–326. Tokyo: Yoshikawa kōbunkan, 1997.

———. "Qing Kangxi-Qianlong nianjian yangtong de jinkou yu liutong wenti" 清康熙一乾隆年間洋銅的進口與流通問題 [The Problem of Import and Circulation of Japanese Copper from the Kangxi to the Qianlong reigns]. In *Zhongguo haiyang fazhanshi lunwenji* 中國海洋發展史論文集 [Essays on Chinese Maritime History], vol. 7, part 1, edited by Tang Xiyong 湯熙勇, 93–144. Taipei: Sun Yat-sen Institute for Social Science and Philosophy, Academia Sinica, 1999.

Matsuura Akira 松浦章. *Edo jidai Tōsen ni yoru Nitchū bunka kōryū* 江戸時代唐船による日中文化交流 [*Sino-Japanese cultural interactions via the Edo-period Chinese junk trade*]. Kyoto: Shibunkaku shuppan 思文閣出版, 2007.

———. "Edo jidai Tōsen ni yoru satō yu'nyū to kokunai shōhi no tenkai" 江戸時代唐船による砂糖輸入と国内消費の展開 [The development of sugar imports from the Chinese junk trade and domestic consumption of sugar during the Edo period]. *Higashi Ajia bunka kōshō kenkyū* 東アジア文化交渉研究 [*Studies on East Asian Cultural Interactions*], 3 (2010): 335–359.

———. *Kinsei Higashi Ajia kaiiki no bunka kōshō* 近世東アジア海域の文化交渉 [*Cultural Interactions in the Early Modern East Asian Maritime Region*]. Kyoto: Shibunkaku shuppan 思文閣出版, 2010.

———. "18–19 seiki ni okeru Nansei shotō hyōchaku Chūgoku hansen yori mita Shindai kōungyō no ichisokumen" 十八〜十九世紀における南西諸島漂着中国帆船より見た清代航運業の一側面 [The 18th–19th century Qing shipping industry as seen through shipwrecked Chinese junks in the Nansei Islands]. *Kansai daigaku tōzai gakujutsu kenkyūjo kiyō* 関西大学東西学術研究所紀要, 16 (Jan 1983): 17–75.

———. *Shindai kaigai bōekishi no kenkyū* 清代海外貿易史の研究 [*Studies on Foreign Trade during the Qing Era*]. Kyoto: Hōyū shoten 朋友書店, 2002.

———. "Nagasaki bōeki ni okeru Kōsetsushō to Binshō" 長崎貿易における江浙商と閩商 [Jiangsu, Zhejiang, and Fujian Merchants in the Nagasaki Trade]. *Shisen* 史泉 (published by Kansai University) 42 (1971): 47–62.

———. "Nisshin bōeki ni yoru tawaramono no Chūgoku ryūnyū ni tsuite" 日清貿易による俵物の中国流入について [On the influx of baled goods into China via trade between Japan and the Qing]. *Senriyama bungaku ronshū* 千里山文学論集, 7 (1972): 19–38.

———. *Shindai kaigai bōekishi no kenkyū* 清代海外貿易史の研究 [Studies on the history of overseas trade during the Qing Dynasty]. Kyoto: Hōyū shoten 朋友書店, 2002.

———. "Shindai kanshō saiben yōdō benkai senseki" 清代官商採辦洋銅辦回船隻 [Qing Official Commercial Ships for the Purchase of Japanese Copper and the Volume of Copper Purchased]. *Kansai daigaku bungaku ronshū* 関西大学文学論集 [Journal of the School of Humanities and Management of Kansai University] 43, no. 4 (1994): 35–56.

———. "Shindai zenki no Sekkō kaikan to kaijō bōeki" 清代前期の浙江海關と海上貿易[Zhejiang Customs and Maritime Trade in the Early Qing] *Shisen* 史泉 85 (1997): 19–32.

Mitsui Rokurō 満井録郎 and Doi Shinichirō 土井進一郎. *Shin Nagasaki nenpyō* 新長崎年表 [New Chronological table of Nagasaki], 2 vols. Nagasaki: Nagasaki bunkensha, 1974–1976.

Momose Hiroshi 百瀬弘. "Chifusokusai sōsho" 知不足齋叢書 (*Zhibuzuzhai-congshu*). In: *Ajia rekishi jiten* アジア歴史辞典 [Dictionary of Asian history] 6: 157. Tokyo: Heibonsha 平凡社 1960.

Murakami Ryū 村上隆. *Kin gin dō no Nihonshi* 金・銀・銅の日本史 [Japanese history of gold, silver, and copper]. Tokyo: Iwanami shoten 岩波書店, 2007.

Nagahara Keiji 永原慶二 and Yamaguchi Keiji 山口啓二, ed. *Kōza Nihon gijutsu no shakaishi (Dai go kan): Saikō to yakin* 講座・日本技術の社会史（第五巻）採鉱と冶金 [Course in Social History of Japanese Technology (Volume 5): Mining and Metallurgy]. Tokyo: Nippon hyōronsha 日本評論社, 1985.

Nagasawa Kikuya 長沢規矩也, ed. *Wakokubon kanshi shūsei* 和刻本漢詩集成 [Collection of Chinese poems using the Japanese method of woodcut printing] vol. 20. Tokyo: Kyūko shoin 汲古書院, 1977.

Nagase-Reimer, Keiko. "Water Drainage in the Mines in Tokugawa Japan: Technological Improvements and Economic Limitations" In *Mining, Monies and Culture in Early Modern Societies*, edited by Nanny Kim and Keiko Nagase-Reimer, 25–42. Leiden: Brill, 2013.

Nagazumi Yōko 永積洋子 "Osaka dōza" 大坂銅座 [The Copper Administration Bureau in Osaka]. In *Nihon sangyō-shi taikei* (dai rokkan) Kinki-chihō hen 日本産業史大系（第六巻）近畿地方篇 [Overview of the industrial history of Japan, vol. 6, the Kinki region], edited by Chihō-shi kenkyū kyōgikai 地方史研究協議会, 408–419. Tokyo: Tōkyō daigaku shuppankai 東京大学出版会, 1960.

———. *Tōsen yushutsunyūhin sūryō ichiran* 1637–1833 *nen—fukugen tōsen kamotsu aratamechō/kihan nimotsu kaiwatari chō* 唐船輸出入品数量一覧 1637–1833 年一復元唐船貨物改帳・帰帆荷物買渡帳—[A Catalogue of Imports and Exports on Chinese Ships, 1637–1833: Restoration of *Tōsen kamotsu aratamechō* and *Kihan nimotsu kaiwatashichō*]. Tokyo: Sōbunsha 創文社, 1987.

Nakajima Satoshi 中嶋敏. "Shinchō no dōsei ni okeru yōdō to tendō" 清朝の銅政における洋銅と滇銅 [Japanese Copper and Yunnan Copper in the Copper Policy of the Qing Dynasty] In *Tōyōshigaku ronshū* 東洋史学論集 [Collected Essays on Oriental History], edited by Nakajima Satoshi 中嶋敏, 161–77. Tokyo: Kyūko shoyin 汲古書院, 1988.

Nakamura Tadashi 中村質. *Kinsei Nagasaki bōeki shi no kenkyū* 近世長崎貿易史の研究 [The History of foreign trade in Nagasaki during the early modern period]. Tokyo: Yoshikawa kōbunkan 吉川弘文館, 1988.

Nihon ginkō chōsakyoku 日本銀行調査局 [Research Center of the Bank of Japan], ed. *Zuroku Nihon no kahei* 図録日本の貨幣 [Japanese Currency—an illustrated record] vol. 3. Tokyo: Tōyō keizai shinpōsha 東洋経済新報社, 1974.

Nihon kagakushi gakkai 日本科学史学会. *Saikō yakin gijutsu* 採鉱冶金技術 [Mining and metallurgical technologies]. Tokyo: Daiichi hōki shuppan 第一法規出版, 1966.

Naikaku bunko 内閣文庫 [Cabinet *Library*]. *Kaitei naikaku bunko kanseki bunrui mokuroku* 改訂内閣文庫漢籍分類目録 [Cabinet *Library* catalog of classical Chinese literature]. Tokyo: Naikaku bunko, 1971.

Ōba Osamu 大庭脩. *Edo jidai ni okeru Tōsen mochiwatashisho no kenkyū* 江戸時代における唐船持渡書の研究 [Research on books traded by Chinese vessels during the Edo period]. Osaka: Kansai Daigaku 関西大学, 1967.

———. *Edo jidai ni okeru Chūgoku bunka juyō no kenkyū* 江戸時代における中国文化受容の研究 [Research related to the reception of Chinese culture during the Edo period]. Kyoto: Dōhōsha 同朋舎, 1984.

———, ed. *Hōreki sannen Hachijōjima hyōchaku Nankinsen shiryō—Edo jidai hyōchaku tōsen shiryōshū ichi* 宝暦三年八丈島漂着南京船資料ー江戸時代漂着唐船資料集 1 [Materials on the wreckage of Nanjing ships at Hachijō-jima in 1753—Collected Edo-period materials on wrecked Chinese ships, vol. 1]. Osaka: Kansai Daigaku shuppanbu 関西大学出版部, 1985.

Ogawa Kuniharu 小川国治. *Edo bakufu yushutsu kaisanbutsu no kenkyū—tawaramono no seisan to shūka kikō* 江戸幕府輸出海産物の研究ー俵物の生産と集荷機構 [Studies on Marine Products Exported by the Edo Bakufu: Structures of Production and Collection of Dried Goods]. Tokyo: Yoshikawa kōbunkan 吉川弘文館, 1973.

Ōno Mizuo 大野瑞男. *Edo bakufu zaisei-shi ron* 江戸幕府財政史論 [History of the financial affairs of the Tokugawa Shogunate]. Tokyo: Yoshikawa kōbunkan 吉川弘文館, 1996.

Ōta Katsuya 太田勝也. "Dōshiromonogae bōeki no sōshi ni tsuite" 銅代物替貿易の創始について [On the beginnings of the copper trade]. *Chūō daigaku daigakuin kenkyū nenpō sōkangō* 中央大学大学院研究年報　創刊号 [The research yearbook of Chūō University] 1 (1971).

———. "Dōza 銅座 [Copper Agency]." In *Kinsei-shi handobukku* 近世史ハンドブック [Handbook of premodern history], edited by Kodama Kōta 児玉幸多 et al. (eds.). Tokyo: Kondō shuppansha 近藤出版社: 157–158, 1972.

Peng Yuxin 彭雨新. "Qingdai qianqi Yunnan tongkuangye ji qi shengchan xingzhi de chutan" 清代前期雲南銅礦業及其生產性質的初探 [A Preliminary Exploration of the Early Qing Era Yunnan Copper Mining Industry and Its Production Mode]. *Wuhan daxue xuebao* 武漢大學學報 [Journal of Wuhan University] 5 (1984): 78–86.

Qiu Lijuan 邱麗娟. "Qingdai Yunnan tongkuang de jingying" 清代雲南銅礦的經營 [Management of Yunnan Copper Mines in the Qing Dynasty]. Unpublished MA Thesis at National Taiwan Normal University, 1991.

Rai Tsutomu 頼惟勤. "Ōkubo Shibutsu 大窪詩仏 (1767–1837)". In *Kokushi daijiten* 国史大辞典 [Dictionary of Japanese history] vol. 2. Tokyo: Yoshikawa kōbunkan 吉川弘文館, 1980, 543.

Sasaki Jun'nosuke 佐々木潤之介. "Kinsei sandō seisaku ni kansuru ichikōsatsu" 近世産銅政策に関する一考察 [A Study of Early Modern Copper Production Policy]. *Shigaku zasshi* 史学雑誌 [History Magazine] 66 (1957): 1–23.

———. "Osaka dōton'ya Osakaya ni tsuite no oboegaki" 大坂銅問屋・大坂屋についての覚書 [Memorandum on Osaka Copper Merchant House Osakaya]. *Kenkyū to hyōron* 研究と評論 [Research and Criticism] 3 (1959): 1–18.

———. "Akita Ani dōzan no keiei: Kansei kaikaku o chūshin ni" 秋田阿仁銅山の経営ー寛政改革を中心に [The Management of the Akita Ani Copper Mine: With a Focus on the Kansei Reform]. Nihon sangyōshi taikei Tōhoku chihō hen 日本産業史大系　東北地方編 [A Survey of Japanese Industrial History: Tōhoku Area], 218–34. Tokyo daigaku shuppankai 東京大学出版会, 1960.

———. *Zairai gijutsu no hakken to kinsei shakai. Gijutsu no shakai-shi dai 2 kan* 在来技術の発見と近世社会　技術の社会史　第 2 巻 [Premodern Society and the Discovery of pre-existing Engineering. Social History of Engineering Vol. 2]. Tokyo: Yūhikaku 有斐閣, 1983.

———. "Dōzan no keiei to gijutsu" 銅山の経営と技術 [Administration and technology of copper Mines]. In Nagahara Keiji 永原慶二 and Yamaguchi Keiji 山口啓二 (eds.) *Kōza Nihon gijutsu no shakaishi (Dai go kan): Saikō to yakin* 講座・日本技術の社会史（第五巻）採鉱と冶金 [Course in the Social History of Japanese Technology (volume 5): Mining and Metallurgy], 177–210. First edition in 1983. Tokyo: Nippon hyōronsha 日本評論社, 1985.

Saeki Tomi 佐伯富. "Shindai Yōseichō ni okeru tsūka mondai" 清代雍正朝における通貨問題 [Currency Problems in the Yongzheng Reign of the Qing Dynasty].

Tōyōshi kenkyū 東洋史研究 [The Journal of Oriental Research)] 18, no.3 (1959): 142–211.

Sekino Tadakazu 関野唯一. *Sekai tōgyō bunka shi* 世界糖業文化史 [A global cultural history of sugar]. Tokyo: Hōkō shobō 邦光書房, 1955.

Shakespeare, William. "Hamlet," in *Shakespeare*, The Viking Portable Library, 1–129. New York: The Viking Press, 1944.

Shimada Ryuto. *The Intra-Asian Trade In Japanese Copper by the Dutch East India Company during the Eighteenth Century*. Leiden: Brill, 2006.

——— 島田竜登. "Kinsei Nihon no dōyushutsu sakugen to chōsen no senkō 近世日本の銅輸出削減と朝鮮の銭荒 [The reduction of copper exports in modern Japan and Korea under the *senkō* policy]". In *Ajia taiheiyō keizaiken-shi 1500–2000* アジア太平洋経済圏史 1500–2000 [Regional economic history of the Asian Pacific from 1500 to 2000], edited by Shimada Ryūto and Kawakatsu Heita 川勝平太, 27–46. Tokyo: Fujiwara shoten 藤原書店, 2003.

Shōji Mitsuo 庄司三男. "Orandajin to Izumiya to Osaka dōfukisho—shōkanchō Edo sanpu ryokō nikki オランダ人と泉屋と大坂銅吹所一商館長江戸参府旅行日記 [The Dutchman, Izumiya and the copper refineries in Osaka—travel journal of the head of the Dutch factory about his journey to Edo]". In *Sakoku Nihon to kokusai kōryū ge* 鎖国日本と国際交流下 [Japan under the so-called seclusion policy and international exchange] vol. 2, edited by Yanai Kenji 箭内 健次, 55–96. Tokyo: Yoshikawa kōbunkan 吉川弘文館, 1988.

Sueoka Teruaki 末岡照啓. "Kinsei zen, chūki ni okeru Sumitomo no keiei kōzō (1)—Toshi shōgyō seiritsu, hatten-ki no kinsei sandō shihon—近世前・中期における住友の経営構造（一)一都市商業成立・発展期の近世産銅資本—[The management structure of Sumitomo in the first half of the early modern period 1: Capital from medieval copper production for the set-up and the early development of its city branch]". *Sumitomo shiryōkanpō* 住友史料館報 No. 24 (1993): 1–89.

[Sumitomo iinkai] Sumitomo kinzoku kōzan kabushiki gaisha Sumitomo Besshi kōzanshi henshū iinkai 住友金属鉱山株式会社住友別子鉱山編集委員会 [Sumitomo Besshi Mine Historical Editing Committee of Sumitomo Metal Mining Co., Ltd.], ed. *Sumitomo Besshi kōzanshi bekkan* 住友別子鉱山史別巻 [History of the Sumitomo Besshi mine, supplement]. Tokyo: Sumitomo kinzoku kōzan kabushikigaisha 住友金属鉱山株式会社, 1991.

———. *Sumitomo Besshi kōzanshi* 住友別子鉱山史 [History of the Sumitomo Besshi mine], vol. 1 (jō 上) Kyoto: Shibunkaku shuppan 思文閣出版, 1991.

Sumitomo shūshishitsu 住友修史室, ed. *Dai ichi ji dōza to Sumitomo: Dōbōeki to bakufu no dōseisaku* 第一次銅座と住友：銅貿易と幕府の銅政策 [The First Copper Agency and Sumitomo: Copper trade and the Bakufu's Copper Policy]. Sen'oku sōkō 泉屋叢考 vol. 18, written by Kobata Atsushi 小葉田淳. Kyoto: Sumitomo shūshishitsu 住友修史室, 1989.

———. *Kinsei Sumitomo no fukisho no kenkyū* 近世住友の吹所の研究 [Research on the Sumitomo refinery in the early modern era] Sen'oku sōkō 泉屋叢考 vol. 19, written by Imai Noriko 今井　典子. Kyoto: Sumitomo shūshishitsu 住友修史室, 1980.

———. *Kinsei zenki ni okeru dō-bōeki to Sumitomo* 近世前期に於ける銅貿易と住友 [Sumitomo and the copper trade in the first part of the early modern era], Sen'oku sōkō 泉屋叢考 vol. 9. Kyoto: Sumitomo shūshishitsu, 1957.

Sun E-tu Zen 任以都. "The Copper Industry of Yunnan: An Historical Sketch." In E-tu Zen Sun 任以都, *Zhongguo jingji shi lunwen xuanji* 中國經濟史論文選集 [Selected Essays on Chinese Economic History], 55–79. Taipei: Xuesheng shuju 學生書局, 1981.

———. "Ch'ing Government and the Mineral Industries before 1800." In E-tu Zen Sun 任以都, *Zhongguo jingji shi lunwen xuanji* 中國經濟史論文選集 [Selected Essays on Chinese Economic History], 119–29. Taipei: Xuesheng shuju 學生書局, 1981.

———. "The Transportation of Yunnan Copper to Peking in the Ch'ing Period." In E-tu Zen Sun 任以都, *Zhongguo jingji shi lunwen xuanji* 中國經濟史論文選集 [Selected Essays on Chinese Economic History], 131–47. Taipei: Xuesheng shuju 學生書局, 1981.

Susuki Shōhei 須々木庄平. *Dōjimamai shijō-shi* 堂島米市場史 [A History of the Rice Market in Dōjima]. Tokyo: Nihon hyōronsha 日本評論社, 1940.

Suzuki Yasuko 鈴木康子. *Kinsei Nichiran bōekishi no kenkyū* 近世日蘭貿易史の研究 [Research on the history of early modern Japanese-Dutch trade]. Kyoto: Shibunkaku shuppan 思文閣出版, 2004.

Tanaka Kenji 田中謙二 and Matsuura Akira 松浦章, eds. *Bunsei kyūnen Enshū hyōchaku Detai-sen shiryō* 文政九年遠州漂着得泰船資料 [Materials on the wreck of the *Detai* in Enshū, 1826]. Osaka: Kansai daigaku shuppanbu 関西大学出版部, 1986.

Tashiro Kazui. "Foreign Relations during the Edo Period: Sakoku Reexamined," translated by Susan Downing Videen. *Journal of Japanese Studies* 8: 2 (1982): 283–306.

——— 田代和生. "Tsushima-han no Chōsen yushutsu dō chōtatsu ni tsuite—bakufu no dō-tōsei to Nissen dō-bōeki no suitai" 対馬藩の朝鮮輸出銅調達について-幕府の銅統制と日鮮銅貿易の衰退—[On the procurement of copper for export to Korea through the Tsushima domain—bakufu's control of copper and the decline of Japanese-Korean copper trade]. *Chōsen gakuhō* 朝鮮学報 66 (1973): 141–208.

Toby, Ronald ロナルド・トビ. *"Sakoku" to iu gaikō* 鎖国」という外交 [Foreign diplomacy called "sakoku"]. Tokyo: Shōgakukan 小学館, 2008.

Ueda Kōichi 植田晃一. "Edo-jidai no kin, gin, dō seiren gijutsu—dai ikkai Bunken to gijustu no gaiyō 江戸時代の金・銀・銅製錬技術　第１回　文献と技術の概要 [The smelting technique for gold, silver and copper in the Edo period. Part 1:

Overview of technology and documents]". *Nihon kōgyōshi kenkyū* 日本鉱業史研究 No. 30 (1995): 55–59.

Wakamatsu Masashi 若松正志. "Kinsei chūkōki no Nagasaki dōbōeki to kokunai sangyō" 近世中後期の長崎銅貿易と国内産業 [Nagasaki copper trade and domestic industry from the middle to the later part of the early modern period]. In Katagiri Kazuo 片桐一男 (ed.), *Nichiran kōryūshi sono hito, mono, jōhō* 日蘭交流史 その人・物・情報 [A History of Japanese Dutch exchange—people, goods, and information]. Kyoto: Shibunkaku shuppan 思文閣出版: 2002, 290–305.

Wang Meng 王萌. "Kangxicao houqi de tongzheng gaige yu Neiwufu guanshang" 康熙朝後期的銅政改革與內務府官商 [Copper Policy Reform and the Imperial Household Department merchants in the Later Years of Kangxi's Reign]. *Qingshi yanjiu* 清史研究 [The Qing History Journal] 1 (2010): 61–72.

Wei Qingyuan 韋慶遠 and Wu Qiyan 吳奇衍. "Qingdai zhuming huangshang Fan shi de xingshuai" 清代著名皇商范氏的興衰 [The Rise and Fall of the Famous Imperial Merchants, the Fan Family of the Qing Era]. *Lishi yanjiu* 歷史研究 [Historical Studies] 3 (1981): 127–44.

Xu Dixin 許滌新 and Wu Chengmin 吳承明, eds. *Zhongguo zibenzhuyi fazhan shi* 中國資本主義發展史 [The History of the Development of Capitalism in China], vol. I. Beijing: Renmin chubanshe, 1985.

Yamaguchi Keiji 山口啓二. "Akita-han seiritsu-ki no hanzaisei" 秋田藩成立期の藩財政 [The Finances of the Akita Domain at the Time of Its Establishment]. In Yamaguchi Keiji 山口啓二, *Bakuhansei seiritsushi no kenkyū* 幕藩制成立史の研究 [Studies of the History of the Establishment of the Shogunate System], 118–48. Kōsō shobō 校倉書房, 1974.

———. "Kingin-zan no gijutsu to shakai" 金銀山の技術と社 会 [Society and technology in gold and silver mines]. In Nagahara and Yamaguchi, *Kōza Nihon gijutsu no shakaishi* 講座・日本技術の社会史 [Social history of Japanese technology], Saikō to yakin 5 採鉱と冶金 5 [Mining and metallurgy 5], edited by Nagahara Keiji 永原慶二 and Yamaguchi Keiji 山口啓二, 141–176. Tokyo: Nippon hyōronsha 日本評論社, 1985.

Yamawaki Teijirō 山脇悌二郎. *Kinsei Nitchū bōeki-shi no kenkyū* 近世日中貿易史の研究 [A Study of Early Modern Sino-Japanese Trade]. Tokyo: Yoshikawa kōbunkan 吉川弘文館, 1960.

———. *Nagasaki no Tōjin bōeki* 長崎の唐人貿易 [Chinese trade in Nagasaki]. Tokyo: Yoshikawa kōbunkan 吉川弘文館, 1964.

Yan Zhongping 嚴中平. *Qingdai Yunnan tongzheng kao* 清代雲南銅政考 [A Study of Copper Administration in Yunnan in the Qing Dynasty]. Beijing: Zhonghua shuju, 1957.

Yang Shouchuan 楊壽川. "Zhang Yunsui yu Qingdai qianqi Yunnan shehui jingji de fazhan" 張允隨與清代前期雲南社會經濟的發展 [Zhang Yunsui and the

Socio-Economic Development of Yunnan in the Early Qing Period]. *Yunnan shehui kexue* 雲南社會科學 [Yunnan Social Science] 4 (1986), 20: 42–47.

Yasukuni Ryōichi 安国良一. "Kinsei Besshi dōzan no shūshi kōzō" 近世別子銅山の収支構造 [The structure of income and expenditure of the Besshi copper mine]. *Sumitomo shiryōkanpō* 住友史料館報 31 (2000): 1–38.

———. "Besshi dōzan no son'eki to sen'oku Osaka honten 別子銅山の損益と泉屋大坂本店 [The profitability of the Besshi copper mine and the main branch of Senoku in Osaka]". *Sumitomo shiryōkanpō* 住友史料館報 No. 33 (2002): 39–96.

———. "Besshi dōzan to Tenpō, Kaei-ki no keiei kiki" 別子銅山と天保・嘉永期の経営危機 [The Besshi copper mine and the management crisis of the Tenpō and Kaei era]. In *Sumitomo shiryōkanpō* 住友史料館報 34 (2003): 43–89.

Yayoshi Mitsunaga 彌吉光長. *Mikan shiryō ni yoru Nihon shuppan bunka dainikan Osaka no honya to tōhon no yunyū* 未刊史料による日本出版文化第二巻 大坂の本屋と唐本の輸入 [Japanese publishing culture from unpublished historical documents vol. 2: Osakan book shops and the import of Chinese books]. Tokyo, Yumani Shobō ゆまに書房, 1988.

Yoon, Byung Nam. *Domain and Bakufu in Tokugawa Japan: The Copper Trade and Development of Akita Domain Mines*. PhD diss., Princeton University, 1995.

Index of Names

Index of Places

www.ingramcontent.com/pod-product-compliance
Lightning Source LLC
LaVergne TN
LVHW010626300926
849162LV00001B/76